计算机系列教材

刘兆毓　郑家农　闫金平　刘华群　武华　编著

计算机英语实用教程
（第五版）

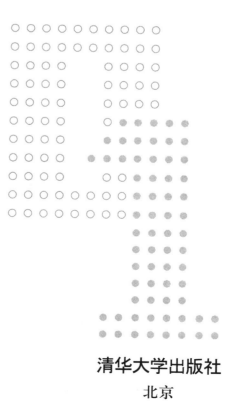

清华大学出版社
北京

内 容 简 介

本书是《计算机英语（第五版）》的精简版本，除了内容精简之外，全书仍按计算机知识结构的层次编写。内容涵盖以下三部分：计算机及计算机网络硬件结构（第1章和第3章），计算机软件（第2、6章为系统软件，第7～11章为应用软件）和因特网应用（第4、5章）。在具体内容的选取上，除了基础知识以外，尽量选择最先进的技术和知识，包括云计算、大数据、3D打印技术等。鉴于因特网应用发展异常迅猛，本书有13课（约占全书38课的1/3）涉及因特网应用。

为便于读者阅读，书中对一些较难理解和翻译的句子做了注释，对涉及的专业术语也都做了解释。

本书主要用作大学计算机及相关专业本科生、专科生的教材，也可供研究生及参加计算机水平考试的考生使用。

本书封面贴有清华大学出版社防伪标签，无标签者不得销售。
版权所有，侵权必究。举报：010-62782989，beiqinquan@tup.tsinghua.edu.cn。

图书在版编目（CIP）数据

计算机英语实用教程/刘兆毓等编著. —5版. —北京：清华大学出版社，2019（2024.8重印）
（计算机系列教材）
ISBN 978-7-302-51697-2

Ⅰ. ①计… Ⅱ. ①刘… Ⅲ. ①电子计算机－英语－高等学校－教材 Ⅳ. ①TP3

中国版本图书馆CIP数据核字（2018）第265414号

责任编辑：袁勤勇　杨　枫
封面设计：常雪影
责任校对：时翠兰
责任印制：刘　菲

出版发行：清华大学出版社
网　　址：https://www.tup.com.cn, https://www.wqxuetang.com
地　　址：北京清华大学学研大厦A座　　邮　编：100084
社　总　机：010-83470000　　邮　购：010-62786544
投稿与读者服务：010-62776969，c-service@tup.tsinghua.edu.cn
质量反馈：010-62772015，zhiliang@tup.tsinghua.edu.cn
课件下载：https://www.tup.com.cn, 010-62795954

印装者：天津安泰印刷有限公司
经　　销：全国新华书店
开　　本：185mm×260mm　　印　张：15.25　　字　数：344千字
版　　次：2003年9月第1版　　2019年5月第5版　　印　次：2024年8月第12次印刷
印　　数：25501～27500
定　　价：39.00元

产品编号：080920-01

前　言

《计算机英语实用教程（第四版）》出版八年以来，计算机和因特网又出现了很多新技术，为了适应这种新的形势，编写了第五版。第五版只保留了第四版中计算机和网络的基础内容，超过 80%是新内容。考虑到长期使用本书的教师和读者的用书习惯，本书编写格式与第四版保持一致，即课文中有注释、关键词和练习题，书后附有习题答案和参考译文。

本书由刘兆毓、郑家农等编著。全书共 11 章，分为 31 节（内含 38 课），其中郑家农编写了第 9~11 章，闫金平编写了第 1、3 章，刘华群编写了第 2、6 章，武华编写了第 4、8 章，刘兆毓编写了第 5、7 章。

本书在编写过程中引用了刘艺、王春生编写的《计算机英语（第 4 版）》的部分内容，在此表示感谢。

由于作者水平有限，书中难免有不当之处，敬请批评指正。

编者
2019 年 1 月于北京

目 录

PART I COMPUTER ARCHITECTURE AND COMPUTER NETWORK

CHAPTER 1 COMPUTER ORGANIZATION AND UNITS ... 3
- 1.1 COMPUTER ORGANIZATION ... 3
- 1.2 MICROPROCESSOR AND SYSTEM BOARD ... 10
- 1.3 MEMORY ... 15
- 1.4 SECONDARY STORAGES ... 19
- 1.5 INPUT AND OUTPUT DEVICES ... 22

CHAPTER 2 SYSTEM SOFTWARE ... 27
- 2.1 WINDOWS 10 ... 27
- 2.2 UNIX AND LINUX ... 32
- 2.3 ANDROID ... 37

CHAPTER 3 COMPUTER NETWORK ... 42
- 3.1 LOCAL AREA NETWORKS (LANs) ... 42
  - 3.1.1 ETHERNET ... 42
  - 3.1.2 WI-FI AND BLUETOOTH ... 46
- 3.2 THE INTERNET ... 51

PART II INTERNET APPLICATIONS

CHAPTER 4 TRADITIONAL INTERNET APPLICATIONS ... 61
- 4.1 OVERVIEW OF WORLD WIDE WEB (WWW) ... 61
  - 4.1.1 ABOUT WWW ... 61
  - 4.1.2 SEARCH ENGINES ... 65
- 4.2 E-MAIL ... 67
- 4.3 ELECTRONIC COMMERCE AND THE INTERNET OF THINGS ... 70
  - 4.3.1 ELECTRONIC COMMERCE ... 70
  - 4.3.2 INTERNET OF THINGS ... 74

CHAPTER 5 NEW INTERNET APPLICATIONS 79
5.1 INSTANT MESSAGING (IM) 79
5.1.1 QQ 79
5.1.2 FACEBOOK 82
5.1.3 TWITTER 85
5.1.4 WECHAT 87
5.2 SOCIAL NETWORKING SERVICE (SNS) 90
5.2.1 WIKI 90
5.2.2 BLOG AND MICROBLOG 92
5.3 CLOUD COMPUTING 96
5.4 BIG DATA 104

PART III PROGRAMMING LANGUAGES AND DATABASES

CHAPTER 6 PROGRAMMING LANGUAGES 113
6.1 C, C++, AND C# 113
6.2 JAVA 117
6.3 MARKUP AND SCRIPTING LANGUAGES 121

CHAPTER 7 DATABASE 128
7.1 DATABASE CONCEPTS 128
7.2 THE WEB AND DATABASES 133

PART IV APPLICATION SOFTWARE

CHAPTER 8 OFFICE AUTOMATION SOFTWARE 141
8.1 THE BASICS OF OFFICE AUTOMATION SOFTWARE 141
8.2 MICROSOFT OFFICE 2013 147

CHAPTER 9 MULTIMEDIA 152
9.1 MULTIMEDIA AND ITS MAJOR CHARACTERISTICS 152
9.2 USAGE/APPLICATION 155

CHAPTER 10 COMPUTER GRAPHICS AND IMAGES 161
10.1 THE VARIOUS COMPUTER GRAPHICS 161
10.2 GRAPHICS SOFTWARE (1) 165
10.3 GRAPHICS SOFTWARE (2) 168

CHAPTER 11　MODERN INDUSTRIAL AUTOMATION 172
11.1　USE OF CAD, CAM, AND CAE 172
11.2　3D PRINTING 175

ANSWERS TO THE EXERCISES 182

参考译文 185
第一部分　计算机体系结构和计算机网络 185
第1章　计算机组成和部件 185
1.1　计算机组成 185
1.2　微处理器和主板 187
1.3　存储器 188
1.4　二级存储器 190
1.5　输入与输出设备 191
第2章　系统软件 192
2.1　Windows 10 192
2.2　UNIX 和 Linux 194
2.3　安卓操作系统 195
第3章　计算机网络 197
3.1　局域网 197
3.2　因特网 199
第二部分　因特网应用 201
第4章　传统因特网应用 201
4.1　万维网概述 201
4.2　电子邮件 203
4.3　电子商务和物联网 204
第5章　因特网新应用 206
5.1　即时消息 206
5.2　社交网络服务 209
5.3　云计算 211
5.4　大数据 213
第三部分　程序设计语言和数据库 215
第6章　程序设计语言 215
6.1　C、C++和 C# 215
6.2　Java 217
6.3　标记和脚本语言 218

第 7 章　数据库ᅟᅟᅟᅟᅟᅟᅟᅟᅟᅟᅟᅟᅟᅟᅟᅟᅟᅟᅟᅟᅟᅟᅟᅟᅟᅟᅟᅟᅟᅟ220
　　7.1　数据库的概念ᅟᅟᅟᅟᅟᅟᅟᅟᅟᅟᅟᅟᅟᅟᅟᅟᅟᅟᅟᅟᅟᅟᅟ220
　　7.2　万维网与数据库ᅟᅟᅟᅟᅟᅟᅟᅟᅟᅟᅟᅟᅟᅟᅟᅟᅟᅟᅟᅟᅟ221

第四部分　应用软件ᅟᅟᅟᅟᅟᅟᅟᅟᅟᅟᅟᅟᅟᅟᅟᅟᅟᅟᅟᅟᅟᅟᅟᅟᅟᅟᅟᅟ223

第 8 章　办公自动化软件ᅟᅟᅟᅟᅟᅟᅟᅟᅟᅟᅟᅟᅟᅟᅟᅟᅟᅟᅟᅟᅟᅟᅟᅟ223
　　8.1　办公自动化软件基本知识ᅟᅟᅟᅟᅟᅟᅟᅟᅟᅟᅟᅟᅟᅟᅟ223
　　8.2　微软 Office 2013ᅟᅟᅟᅟᅟᅟᅟᅟᅟᅟᅟᅟᅟᅟᅟᅟᅟᅟᅟᅟᅟᅟ224

第 9 章　多媒体ᅟᅟᅟᅟᅟᅟᅟᅟᅟᅟᅟᅟᅟᅟᅟᅟᅟᅟᅟᅟᅟᅟᅟᅟᅟᅟᅟᅟᅟᅟ226
　　9.1　多媒体及其主要特点ᅟᅟᅟᅟᅟᅟᅟᅟᅟᅟᅟᅟᅟᅟᅟᅟᅟᅟ226
　　9.2　多媒体应用ᅟᅟᅟᅟᅟᅟᅟᅟᅟᅟᅟᅟᅟᅟᅟᅟᅟᅟᅟᅟᅟᅟᅟᅟᅟ227

第 10 章　计算机图形图像ᅟᅟᅟᅟᅟᅟᅟᅟᅟᅟᅟᅟᅟᅟᅟᅟᅟᅟᅟᅟᅟᅟ228
　　10.1　各种各样的计算机图形ᅟᅟᅟᅟᅟᅟᅟᅟᅟᅟᅟᅟᅟᅟᅟᅟ228
　　10.2　图形软件（1）ᅟᅟᅟᅟᅟᅟᅟᅟᅟᅟᅟᅟᅟᅟᅟᅟᅟᅟᅟᅟᅟᅟ229
　　10.3　图形软件（2）ᅟᅟᅟᅟᅟᅟᅟᅟᅟᅟᅟᅟᅟᅟᅟᅟᅟᅟᅟᅟᅟᅟ230

第 11 章　现代工业自动化ᅟᅟᅟᅟᅟᅟᅟᅟᅟᅟᅟᅟᅟᅟᅟᅟᅟᅟᅟᅟᅟᅟ231
　　11.1　CAD、CAM、CAE 的应用ᅟᅟᅟᅟᅟᅟᅟᅟᅟᅟᅟᅟᅟᅟ231
　　11.2　3D 打印ᅟᅟᅟᅟᅟᅟᅟᅟᅟᅟᅟᅟᅟᅟᅟᅟᅟᅟᅟᅟᅟᅟᅟᅟᅟᅟᅟ233

BIBLIOGRAPHYᅟᅟᅟᅟᅟᅟᅟᅟᅟᅟᅟᅟᅟᅟᅟᅟᅟᅟᅟᅟᅟᅟᅟᅟᅟᅟᅟᅟᅟᅟᅟᅟᅟ235

PART I

COMPUTER ARCHITECTURE AND COMPUTER NETWORK

CHAPTER 1
COMPUTER ORGANIZATION AND UNITS

1.1 COMPUTER ORGANIZATION

1. Computer Organization

A computer is a programming, electronic device that accepts input, performs operations or processing on the data, and outputs and stores the results. Because it is programmable, the instructions—called the program—tell the computer what to do. The relationships between these four main computer operations (input, processing, output, and storage) are shown in Figure 1-1.

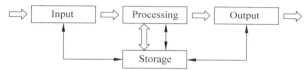

Figure 1-1　Basic operations within a computer

The corresponding devices to perform these tasks are input devices, processing devices, output devices, and storage devices.

(1) Input Devices

An input device is any piece of equipment that supplies materials (input) to the computer. The most common input devices are the keyboard and mouse (See Figure 1-2). Other possibilities include image and bar-code scanners, joysticks, touch screens, digital cameras, electronic pens, fingerprint readers, and microphones. Input devices for a stereo system might be a CD player and antenna.

Figure 1-2　Hardware of a computer system

(2) Processing Unit

The heart of any computer system is the central processing unit (CPU), located inside the computer's main box or system unit.

A processor is composed of two functional units—a control unit and an arithmetic/logic unit—and a set of special workspaces called registers.

Figure 1-3 depicts its structure, in which the Internal CPU Interconnection provides communication among the Control Unit, ALU, and Registers.

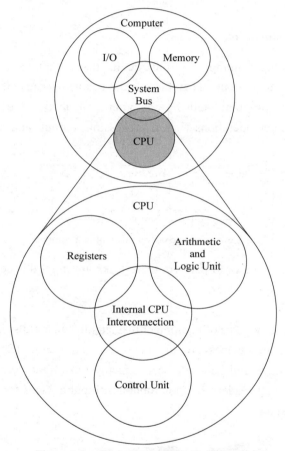

Figure 1-3 Central Processing Unit (CPU)

The control unit is the functional unit that is responsible for supervising the operation of the entire computer system.

The control unit fetches instructions from memory and determines their types or decodes them. It then breaks each instruction into a series of simple small steps or actions. By doing this, it controls the step-by-step operation of the entire computer system.

The arithmetic and logic unit (ALU) is the functional unit that provides the computer with logical and computational capabilities. Data are brought into the ALU by the control unit, and the ALU performs whatever arithmetic or logic operations are required to help carry out the instruction [1].

A register is a storage location inside the processor. Registers in the control unit are used to keep track of the overall status of the program that is running. Control unit registers store information such as the current instruction, the location of the next instruction to be executed, and the operands of the instruction [2]. In the ALU, registers store data items that are added, subtracted, multiplied, divided, and compared. Other registers store the results of arithmetic and logic operations.

(3) Output Devices

Like input units, output devices are instruments of interpretation and communication between humans and computer systems of all sizes. These devices take output results from the CPU in machine-coded form and convert them into a form that can be used (a) by people (e.g. a printed and/or displayed report) or (b) as machine input in another processing cycle [3].

In personal computer systems, display screen and desktop printers are popular output devices. Larger and faster printers, many online workstations, and magnetic tape drives are commonly found in larger systems.

(4) Storage Devices

Storage is a computer section used primarily for storing information such as instructions, programs and data.

There are two types in storage devices, one is the memory (sometimes called as primary storage), another is the secondary storage. Primary storage is located within the system unit that houses the CPU and other components [4]. Secondary storages include the storage media and drives, We will describe them in section 1.4 of this textbook.

An arbitrary desktop computer (not necessarily a PC) is shown in Figure 1-4. It has a large main memory to hold the operating system, applications and data, and an interface to mass storage devices (disks and DVD/CD-ROMs). It has a variety of I/O devices for user input (keyboard, mouse, and audio), user output (display interface and audio), and connectivity (networking and peripherals). The fast processor requires a system manager to monitor its core temperature and supply voltages, and to generate a system reset.

2. Types of Computer

There are four types of computers: supercomputers, mainframe computers, midrange computers, and microcomputers.

① **Supercomputers** are the most powerful type of computer. These machines are special high-capacity computers used by very large organizations. IBM's Blue Gene is considered by many to be the fastest computer in the world.

② **Mainframe computers** occupy specially wired, air-conditioned rooms. Although not nearly as powerful as supercomputers, mainframe computers are capable of great processing speeds and data storage [5]. For example, insurance companies use mainframes to process information about millions of policyholders.

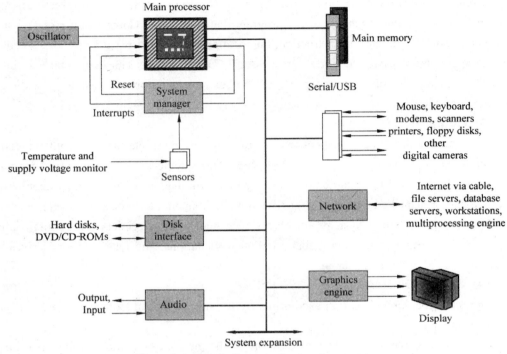

Figure 1-4　Block diagram of a generic computer

③ **Midrange computers** also referred to as servers, are computers with processing capabilities less powerful than a mainframe computer yet more powerful than a microcomputer[6]. Originally used by medium-sized companies or departments of large companies to support their processing needs, today midrange computers are most widely used to support or serve end users for such specific needs as retrieving data from a database or supplying access to application software.

④ **Microcomputers** are the least powerful, yet the most widely used and fastest-growing, type of computer. There are four types of microcomputers: desktop, notebook, tablet PC and handheld computers (See Figure 1-5). Desktop computers are small enough to fit on top of or alongside a desk yet are too big to carry around [7]. Notebook computers, also known as laptop computers, are portable and lightweight and fit into most briefcases. Tablets, also known as tablet computers, are the newest type of computer. They are smaller, lighter, and generally less powerful than notebook. Like a notebook, tablets have a flat screen but typically do not have a standard keyboard. Instead tablets typically use a virtual keyboard that appears on the screen and is touch-sensitive [8]. The best-known tablet is Apple's iPad [9]. Handheld computers are the smallest and are designed to fit into the palm of one hand. These systems contain an entire computer system, including the electronic components, secondary storage, and input and output devices. Personal Digital Assistants (PDAs)[10] and smartphones are the most widely used handheld computers. Smartphones are cell phones with wireless connections to the Internet. Their growth has been explosive in the past few years.

Desktop　　　　　　　Handheld　　　　Tablet PC　　　　　　　Notebook

Figure 1-5　Microcomputers

NOTES

[1] 这是一个 and 连接的并列句。后一个分句中的 whatever 是关系代词，引导后面的宾语从句。

[2] 长句中 such as 引导的同位语中有三个并列的宾语。

[3] 由 and 连接的并列句，后一句中 that 引导的定语从句修饰 form。

[4] that 引导的定语从句，修饰 system unit；house 为动词，原意为"留宿""收容"，此处为"含有""包含"之意。

[5] Although 引导的是让步状语从句。

[6] 连接词 yet 连接的两个句子都是比较句型。

[7] 连接词 yet 连接的两个句子，后一句省略了主语 desktop computers。

[8] that appears…and is 是定语从句，修饰 keyboard。

[9] iPad 是由苹果公司于 2010 年开始发布的平板电脑系列，定位介于苹果公司的智能手机 iPhone 和笔记本电脑产品之间（屏幕中有 4 个虚拟程序固定栏）。与 iPhone 布局一样，iPad 提供浏览互联网、收发电子邮件、观看电子书、播放音频或视频、玩游戏等功能。

[10] PDA 即掌上电脑，主要用于存储和访问个人信息，如地址、电话以及记事等。

KEYWORDS

architecture	体系结构，结构，层次结构，总体结构，结构格式
programmable	可编程的
input	输入
output	输出
store, storage	存储，存储器
bar-code	条形码
scanner	扫描仪
joystick	操纵杆
touch screen	触摸屏
digital camera	数码相机

electronic pen	电子笔
fingerprint reader	指纹阅读器
stereo system	立体音响系统
CD(Compact Disk)	高密度磁盘，光盘，激光唱盘
player	播放器
CPU(Central Processing Unit)	中央处理器
main box	主机箱
control unit	控制器，控制部件
ALU(Arithmetic and Logic Unit)	算术／逻辑部件
register	寄存器
memory	存储器
decode	译码
operand	操作数
communication	通信
machine-coded	机器编码
display screen	显示屏
desktop printer	台式打印机
online	在线，联机
workstation	工作站
magnetic tape drive	磁带驱动器
primarily storage	主存储器
secondary storage	二级存储器
supercomputer	超级计算机
mainframe computer	大型计算机
midrange computer	中型计算机
server	服务器
microcomputer	微型计算机
desktop computer	台式计算机
portable	便携的，手提的，可移植的
screen	屏幕
keyboard	键盘
notebook computer	笔记本电脑
tablet PC	平板电脑
handheld computer	手持计算机
retrieve	检索，使恢复
database	数据库
access	访问，存取，接入，进入
input device	输入设备

output device	输出设备
PDA(Personal Digital Assistants)	个人数字助理
smartphone	智能电话
cell phone	蜂窝电话，移动电话，手机
wireless	无线（的）

EXERCISES

Match the following terms to the appropriate definitions.

1. _____ computer
2. _____ instructions
3. _____ cell phones
4. _____ input device
5. _____ CPU
6. _____ registers
7. _____ control unit
8. _____ the smallest computer
9. _____ output device
10. _____ storage
11. _____ supercomputers
12. _____ mainframe computers
13. _____ servers
14. _____ microcomputers
15. _____ basic operations within a computer
16. _____ ALU

    a. a computer section stored instructions, programs and data

    b. the most widely used and fastest-growing computers

    c. input, processing, storage, and output

    d. supervises the operation of the entire computer system

    e. have great processing speeds and data storage

    f. a programming electronic device

    g. instrument of interpretation and communication between human and computer systems

    h. tell the computer what to do

    i. the heart of any computer system

    j. has logical and computational capabilities

    k. midrange computers

    l. the most powerful type of computer

    m. storage location inside the processor

    n. supplies materials to the computer

o. smartphones

p. handheld

1.2 MICROPROCESSOR AND SYSTEM BOARD

1. Microprocessors

In a microcomputer system, the central processing unit (CPU) or processor is contained on a single chip called the microprocessor. The microprocessor is either mounted onto a carrier package that plugs into the system board or contained within a cartridge that plugs into a special slot on the system board[1] (See Figure1-6). The microprocessor is the "brains" of the computer system. It has two basic components: the control unit and the arithmetic-logic unit. Functions of these two units have been mentioned in section 1.1 of this textbook.

Figure 1-6　Microprocessor carrier package and cartridge

2. Microprocessor chips

Chip capacities are often expressed in word sizes. A word is the number of bits (such as 16, 32, or 64) that can be accessed at one time by the CPU. The more bits in a word, the more powerful—and the faster—the computer is. As mentioned previously, eight bits group together to form a byte. A 32-bit-word computer can access 4 bytes at a time. A 64-bit-word computer can access 8 bytes at a time. Therefore, the computer designed to process 64-bit words is faster.

Older microcomputers typically process data and instructions in millionths of a second or microseconds. Newer microcomputers are much faster and process data and instructions in billionths of a second, or nanoseconds. Supercomputers, by contrast, operate at speeds measured in picoseconds—1000 times as fast as microcomputers (See Figure 1-7).

The two most significant recent developments in microprocessors are the 64-bit processor and the dual-core chip. Until recently, 64-bit processors were only used in large mainframe and supercomputers. All of that is changing as 64-bit processors are becoming commonplace in today's more powerful microcomputers.

CHAPTER 1 COMPUTER ORGANIZATION AND UNITS

Unit	Speed
Microsecond	Millionth of a second
Nanosecond	Billionth of a second
Picosecond	Trillionth of a second

Figure 1-7 Processing speeds

A new type of chip, the dual-core chip, can provide two separate and independent CPUs. These chips allow a single computer to run two programs at the same time. For example, access could be searching a large database while the end user is creating a multimedia presentation with PowerPoint[2]. More significantly, however, is the potential for microcomputers to run very large complex programs that previously were run only on mainframe and supercomputers. This requires specifically designed programs that are divided into parts that each CPU could process independently[3]. This approach is called parallel processing.

3. Multi-core processor

A multi-core processor is a single computing component with two or more independent actual central processing units (called "cores"). Figure 1-8 shows a diagram of a generic dual-core processor.

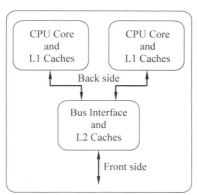

Figure 1-8 A diagram of a generic dual-core processor, with CPU-local
level 1 caches, and a shared, on-die level 2 cache

Manufacturers typically integrate the cores onto a single integrated circuit die (known as a chip multiprocessor or CMP), or onto multiple dies in a single chip package.

Multicore processors may have two cores (dual-core CPUs, for example, AMD Phenom II X2 and Intel Core Duo), four cores (quad-core CPUs, for example, Intel's i7 processors), six cores, eight cores, or more[4].

Multi-core processors are widely used across many application domains including general-purpose, embedded, network, digital signal processing (DSP), and graphics[5].

Commercially, Adapteva Epiphany, a many-core processor architecture which allows up to 4096 processors on-chip, although only a 16 core version has been commercially produced[6].

4. System board

The system board is also known as the motherboard. The system board is the communications medium for the entire computer system. Every component of the system unit connects to the system board. It acts as a data path allowing the various components to communicate with one another. External devices such as the keyboard, mouse, and monitor could not communicate with the system unit without the system board.

On a desktop computer, the system board is located at the bottom of the system unit or along one side. It is a large flat circuit board covered with a variety of different electronic components including sockets, slots, and bus lines (See Figure 1-9).

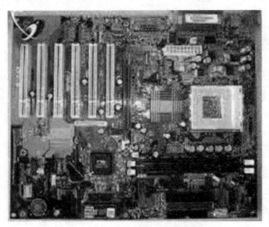

Figure 1-9 System board

① **Sockets** provide a connection point for small specialized electronic parts called chips. Chips consist of tiny circuit boards etched onto squares of sandlike material called silicon.[7] These circuit boards can be smaller than the tip of your finger. A chip is also called a silicon chip, semiconductor, or integrated circuit. Chips are mounted on carrier packages (See Figure 1-10). These packages either plug directly into sockets on the system board or onto cards that are then plugged into slots on the system board. Sockets are used to connect the system board to a variety of different types of chips, including microprocessor and memory chips.

② **Slots** provide a connection point for specialized cards or circuit boards. These cards provide expansion capability for a computer system. For example, a modem card plugs into a slot on the system board to provide a connection to the Internet.

③ **Connecting lines** called bus lines provide pathways that support communication among the various electronic components that are either located on the system board or attached to the system board[8].

Figure 1-10   Chips mounted onto a carrier package

Notebook, tablet PC, and handheld system boards are smaller than desktop system boards. However, they perform the same functions as desktop system boards.

NOTES

[1] 由 either…or…构成的并列句，每个句子里都有由 that 引导的定语从句。

[2] while 连接的是两个并列句子，这一句与上一句有密切关系，故 access 可译为程序。

[3] 句中有两个 that 引导的定语从句。

[4] AMD Phenom Ⅱ X2 是 AMD 公司生产的中高端 CPU，其中文译名为"AMD 羿龙Ⅱ"。Intel Core Duo，即 Core 2 Duo。Intel（英特尔）公司沿用奔腾系列的命名规则，将新系列"酷睿"（Core）芯片命名为 Core 2 Duo。Intel's i7 processors 即酷睿 i7，是由 Intel 生产的面向中高端用户的 CPU 系列标识。

[5] including…，现在分词短语作定语，修饰 domains。

[6] Adapteva 是一个名不见经传的芯片设计小公司，此次 64 核 CPU 采用的是 Epiphany 架构，是一种用来进行 DSP 数字信号处理的芯片架构。

[7] 句中有两个过去分词短语作定语：etched 和 called。

[8] 长句。句中有两个 that 引导的定语从句。

KEYWORDS

microprocessor	微处理器
chip	芯片
system board	系统板，主板
carrier package	承载插件
cartridge	盒
slot	（插）槽

capacity	容量，能力
word sizes	字长，字尺寸
bit	比特，二进制位
byte	字节
microsecond	微秒
nanosecond	毫微秒，纳秒
picosecond	皮秒
core	核，核心，芯子
parallel processing	并行处理
die	电路小片，模片，饼图
DSP(Digital Signal Processing)	数字信号处理
embedded	嵌入式的
motherboard	主板，母板
integrated circuit	集成电路
data path	数据通路（路径）
socket	插座，套接字，网络应用程序接口

EXERCISES

Multiple Choices

1. Microprocessor _____.
    a. is the "brains" of the computer system
    b. is mounted onto a carries package
    c. is contained within a cartridge
    d. has two basic components: the control unit and arithmetic-logic unit
2. We often express chip capacities in word size, word size can be _____.
    a. one byte   b. two bytes   c. three bytes   d. four bytes
3. We usually measure the processing speed in today's computer according to the _____.
    a. thousandth of a second   b. millionth of a second
    c. billionth of a second   d. trillionth of a second
4. The 64-bit processor can be used in _____.
    a. large mainframe computers   b. older microcomputers
    c. supercomputers   d. more powerful microcomputers
5. With dual-core chip a microcomputer can _____.
    a. run two programs at the same time
    b. run very large complex programs
    c. take parallel processing
    d. be used to access a large database while creating a multimedia presentation with PowerPoint

6. Multicore processor may have _____ cores.
   a. one   b. two   c. four   d. sixteen
7. Dual-core processor has _____.
   a. one CPU core   b. two CPU core   c. one-level cache   d. two-level cache
8. Multicore processors can be used for _____ applications.
   a. network   b. DSP   c. graphics   d. embedded
9. System board is _____.
   a. used for desktop computer only
   b. the communications medium for the entire computer system
   c. known as the motherboard
   d. a data path allowing the various components to communicate with one another
10. A connection point on a system board can be provided by _____.
    a. chips   b. slots
    c. electronic components   d. sockets
11. We can call a chip as _____.
    a. silicon chip   b. carrier package
    c. integrated circuit   d. semiconductor
12. Slots can be used for _____.
    a. specialized cards   b. specialized circuit boards
    c. expansion capability of a computer   d. modem card

1.3 MEMORY

1. Memory System Desiderata

The memory system has three desiderata.
① **Size:** infinitely large, no constraints on program or data set size.
② **Speed:** infinitely fast, latency equal to the fastest memory technology available[1].
③ **Cost:** per bit cost should approach the lowest-cost technology available.

Clearly, these specifications cannot all be achieved as they are mutually exclusive. However, with the semiconductor and magnetic memory technology of today, these specifications are closely approximated.

2. Memory

Memory is a holding area for data, instructions, and information. Like microprocessors, memory is contained on chips connected to the system board. There are three well-known types of memory chips: random-access memory (RAM), read-only memory (ROM), and complementary metal-oxide semiconductor (CMOS).

(1) RAM

Random-access memory (RAM) chips hold the program (sequence of instructions) and data that the CPU is presently processing. Before data can be processed or a program can be run, it must be in RAM. For this reason, RAM is sometimes referred to as primary storage. RAM is called temporary or volatile storage because everything in most types of RAM is lost as soon as the microcomputer is turned off[2]. It is also lost if there is a power failure or other disruption of the electric current going to the microcomputer.

There are two types of RAM: Dynamic RAM (DRAM) and Static RAM (SRAM).

① DRAM

Recent advances in DRAM chips have produced three types of DRAM chips:

- Synchronous DRAM (SDRAM) is faster than conventional RAM chips and more expensive. SDRAM chips streamline operations by coordinating or synchronizing the movement of data and instructions between the chip and other components in the system unit [3].
- Double data rate SDRAM (DDR SDRAM), also known as SDRAM II, is faster, more reliable, and more expensive than SDRAM. DDR SDRAM chips are able to transfer twice (double) as much data in the same amount of times as SDRAM.
- Direct DRAM is the fastest and most expensive.

Almost all of today's microcomputers use a combination of DRAM chips.

② SRAM

Static RAM (SRAM), like DRAM, requires a constant supply of power. Compared to DRAM, SRAM does not require as much power, is faster, and is more reliable. SRAM is also more expensive and typically used for specialized applications. One of these applications is for cache memory or RAM cache [4].

③ CACHE

Cache (pronounced "cash") memory improves processing by acting as a temporary high-speed holding area between the memory and the CPU. In a computer with a cache (not all machines have one), the computer detects which information in RAM is most frequently used. It then copies that information into the cache. When needed, the CPU can quickly access the information from the cache.

There are three different types or levels of cache:

- Level 1 (L1), also known as primary cache and internal cache, is built into the microprocessor chip.
- Level 2 (L2), also known as external cache, is slower than L1 but has a greater capacity. In older computers with older microprocessors, the Level 2 cache is located on a chip that is plugged into the system board. New computers with newer microprocessors have the L2 caches built into the microprocessor. This arrangement, sometimes referred to as advanced transfer cache, provides a faster response than

cache located on the system board [5].
- Level 3 (L3), the newest type of cache, works with special microprocessor L2 caches. L3 caches are not built into the microprocessor. Rather, L3 caches use SDRAM chips located on the system board.

Most of today's microcomputers have two or three types of cache. The most powerful have all three types.

④ FLASH RAM

Flash RAM or flash memory chips can retain data even if power is disrupted. This type of RAM is the most expensive and used primarily for special applications such as digital cell telephones, digital video cameras, and portable computers.

(2) ROM

Read-only memory (ROM) chips have programs built into them at the factory. Unlike RAM chips, ROM chips are not volatile and cannot be changed by the user. "Read only" means that the CPU can read or retrieve data and programs written on the ROM chip. However, the computer cannot write—encode or change—the information or instructions in ROM.

ROM chips typically contain special instructions for detailed computer operations. For example, ROM instructions are needed to start a computer, give keyboard keys their special control capabilities, and put characters on the screen.

(3) CMOS

A complementary metal—oxide semiconductor (CMOS) chip provides flexibility and expandability for a computer system. It contains essential information that is required every time the computer system is turned on. The chip supplies such information as the current date and time, amount of RAM, type of keyboard, mouse, monitor, and disk drives. Unlike RAM, it is powered by a battery and does not lose its contents when the power is turned off. Unlike ROM, its contents can be changed to reflect changes in the computer system such as increased RAM and new hardware devices.

NOTES

[1] latency 即等待时间，潜伏时间。此处指存储器访问时间。

[2] 在 because 引导的原因状语从句中，as soon as 引导的是时间状语从句。

[3] streamline 为动词，意为使（工作、生产等）简化而更具效率。

[4] cache memory 就是 cache；RAM cache 一般作为 L2 cache 使用，在苹果计算机中指 disk cache。

[5] transfer cache，传输缓存，其中关键是 transfer，它能控制 cache 和其中的内容。

KEYWORDS

RAM（random-access memory）　　随机存储器

ROM (read-only memory)　　　　只读存储器
CMOS (Complementary Metal-Oxide Semiconductor)
　　　　　　　　　　　　　　　互补金属氧化物半导体
volatile　　　　　　　　　　　易失性
DRAM (Dynamic RAM)　　　　　动态随机存储器
SRAM (Static RAM)　　　　　　静态随机存储器
cache　　　　　　　　　　　　高速缓冲存储器
flash memory　　　　　　　　 快闪存储器，闪存

EXERCISES

Fill in the blanks with appropriate words or phrases found behind this exercises.

1. There are three specifications in the memory system, they are _____.
2. An area that holds data, instructions, and information is called as _____.
3. There are three types of memory chips, they are _____.
4. Which chips are sometimes referred to primary storage? _____.
5. RAM is called _____ storage as everything in most types of RAM is lost as soon as the microcomputer is turned off.
6. There are two types of RAM: _____.
7. There are three types of DRAM: _____.
8. _____ acts as a temporary high-speed holding area between the memory and the CPU.
9. _____ is also known as external cache.
10. _____ is also known as primary cache.
11. _____ is faster than _____.
12. The newest type of cache is _____.
13. The most powerful microcomputers have _____ of cache.
14. _____ can retain data even if power is disrupted.
15. _____ chips are not volatile.
16. _____ are needed to start a computer.
17. CMOS chip is powered by _____.
18. _____ supplies such information as the current date, time and so forth.

　　a. RAM chips
　　b. Cache
　　c. CMOS chip
　　d. Flash RAM
　　e. L3 cache
　　f. size, speed, and cost
　　g. L2 cache
　　h. temporary or volatile

i. RAM, ROM, and CMOS
j. ROM instructions
k. all three types
l. DRAM and SRAM
m. memory
n. a battery
o. L1 cache
p. L1 cache, L2 cache
q. ROM
r. SDRAM, DDR SDRAM and Direct DRAM

1.4 SECONDARY STORAGES

As most RAM provides only temporary or volatile storage, therefore we need more permanent or nonvolatile storage for data and programs. We also need external storage because users need much more storage capacity than is typically available in a computer's primary or RAM memory [1].

Secondary storage provides permanent or nonvolatile storage. Using secondary storage devices such as a hard disk drive, data and programs can be retained after the computer has been shut off. This is accomplished by writing files to and reading files from secondary storage devices.

1. HARD DISKS

Compared to floppy disks, hard disks are able to store and retrieve information much faster and have a greater capacity [2].

There are three types of hard disks:
① **Internal hard disk,** an internal hard disk is located inside the system unit.
② **Hard-disk cartridges,** as easy to remove as a cassette from a videocassette recorder.
③ **Hard-disk packs,** removable storage devices used to store massive amounts of information. Their capacity far exceeds the other types of hard disks.

Three ways to improve the performance of hard disks are disk caching, redundant arrays of inexpensive disks (RAID), and file compression/decompression.

2. OPTICAL DISCS

Today's optical discs can hold over 50 gigabytes of data. That is the equivalent of millions of typewritten pages or a medium-sized library all on a single disc [3].

(1) Compact disc

Compact disc, or as it is better known, CD, is one of the most widely used optical formats.

There are three basic types of CDs: read only, write once, and rewritable.

① **Read only-CD-ROM,** which stands for compact disc-read-only memory, is similar to a commercial music CD. Read only means it cannot be written on or erased by the user.

② **Write once-CD-R**, which stands for CD-recordable, can be written to once. After that they can be read many times without deterioration but cannot be written on or erased.

③ **Rewriteable-CD-RW**, which stands for compact disc rewritable. Also known as erasable optical discs.

(2) Digital versatile disc (DVD)

DVD stands for digital versatile disc or digital video disc. This is a newer format that has replaced CDs as the standard optical disc.

(3) High-definition disc

High-definition disc is the next generation of optical disc and is called as hi def (high definition), with a far greater capacity than DVDs. Like CDs and DVDs, hi def has three basic types: read only, write once, and rewriteable.

3. Other types of secondary storage

(1) Solid-state storage

Each of the secondary storage devices discussed thus far has moving parts. For example, hard disks rotate and read/write heads move in and out. Unlike these devices, solid-state storage devices have no moving parts. Data and information are stored and retrieved electronically directly from these devices much as they would be from conventional computer memory[4].

① **Flash memory cards** are credit card-sized solid-state storage devices widely used in notebook computers.

② **USB drives,** also known as USB flash drives, are so compact that they can be transported on a key ring or a necklace. These drives conveniently connect directly to a computer's USB port to transfer files.

(2) Internet hard drives

Special service sites on the Web provide users with storage. This storage is called an Internet hard drive (See Figure 1-11).

Advantages of Internet hard drives compared to other types of secondary storage include low cost and the flexibility to access information from any location using the Internet.

(3) Magnetic tape

Magnetic tape is a slower sequential access device. Although slower to access specific information, magnetic tape is an effective and commonly used tool for backing up data.

CHAPTER 1　COMPUTER ORGANIZATION AND UNITS

Figure 1-11　An Internet hard drive site

NOTES

[1] more…than…构成的比较句，在 than 后面省略了 that。

[2] floppy disk，软（磁）盘，早期的微型计算机中常用的便携式磁盘存储器，现已很少用。

[3] all on a single disc，介词短语作状语。

[4] much as 后面是由虚拟语气构成的方式状语从句；they 代表 data and information。

KEYWORDS

external storage	外部存储器
storage capacity	存储容量
floppy disk	软（磁）盘
storage medium	存储媒体
hard disk	硬（磁）盘
hard-disk cartridge	盒式硬盘
hard-disk pack	硬盘组
cassette	盒，磁带盒
videocassette recorder	盒式录像机
disk cache	磁盘高速缓存
RAID (Redundant Arrays of Independent Disks)	独立磁盘冗余阵列
file compression	文件压缩
file decompression	文件解压缩
optical disc	光盘

GB (gigabytes)	吉字节
compact disc	高密度磁盘，光盘
CD-ROM（Compact Disc-Read-Only-Memory）	光盘只读存储器
CD-R（CD-Recordable）	可录制光盘
CD-RW（Compact Disc Rewritable）	可重写光盘
DVD（Digital Versatile Disc，Digital Video Disc）	数字通用光盘，数字视频光盘
high-definition discs	高清晰光盘
hi def（high definition）	高清晰度，高分辨率
solid-state storage	固态存储器，半导体存储器
magnetic type	磁带
read/write head	读/写磁头
flash memory card	闪存卡
USB drive	U盘驱动器

EXERCISES

True / False

1. _____ Most RAM is a permanent or non volatile storage.
2. _____ We need external storage for much more storage capacity.
3. _____ Hard disks are permanent or nonvolatile storage.
4. _____ Using RAM, data and programs can be retained after the computer has been shut off.
5. _____ DVD stands for digital versatile disk.
6. _____ There are three types of hard disks.
7. _____ Internal hard disk is a removable disk.
8. _____ Solid-state storage devices have moving parts.
9. _____ External storage provides permanent storage.
10. _____ Internet hard drives are used to serve to sites on the Web.

1.5　INPUT AND OUTPUT DEVICES

1. Input Devices

(1) Keyboard Entry

One of the most common ways to input data is by keyboard.

There are a wide variety of different keyboard designs. They range from the full-sized to miniature and from rigid to flexible [1]. Most common types are:

- Traditional Keyboards(Figure 1-12)
- Ergonomic Keyboards(Figure 1-14)
- PDA keyboards
- Flexible Keyboards(Figure 1-13)
- Wireless keyboards

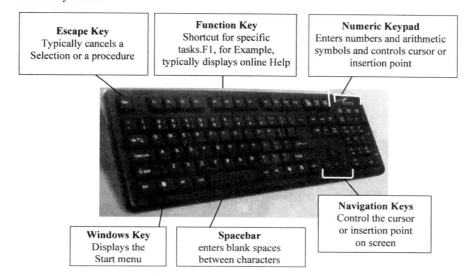

Figure 1-12　Traditional keyboard

Figure 1-13　Flexible keyboard　　　　Figure 1-14　Ergonomic keyboard

(2) Pointing Devices

① Mice [2]

A mouse controls a pointer that is displayed on the monitor. The mouse pointer usually appears in the shape of an arrow. Although there are several different mouse types, there are three basic designs: mechanical mouse, optical mouse, cordless or wireless mouse.

② Joysticks

A joystick is the most popular input device for computer games (See Figure 1-15).

③ Touch Screen

A touch screen is a particular kind of monitor and is commonly used in restaurants, automated teller machines (ATMs), and information centers (See Figure 1-16).

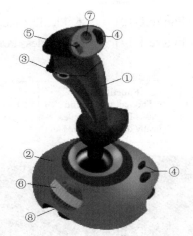

Figure 1-15　Video game joystick elements: 1. stick, 2. base, 3. trigger, 4. extra buttons, 5. autofire switch, 6. throttle, 7. hat switch (POV hat), 8.suction cup

④ Light Pen

A light pen is a light-sensitive penlike device. The light pen is placed against the monitor. This closes a photoelectric circuit and identifies the spot for entering or modifying data. For example, light pens are used to edit digital images and drawings.

⑤ Stylus

A stylus is a penlike device commonly used with tablet PCs and PDAs (See Figure 1-17). A stylus uses pressure to draw images on a screen. A stylus interacts with the computer through handwriting recognition software. Handwriting recognition software translates handwritten notes into a form that the system unit can process.

Figure 1-16　A touch screen: a consumer application　　　　Figure 1-17　Stylus

(3) Scanning Devices

Scanners move across text and images. Scanning devices convert scanned text and images into a form that the system unit can process. There are four types of scanning devices: optical scanners, card readers, bar code readers and character and mark recognition devices.

(4) Image Capturing Devices

Like traditional copy machines, image capturing devices are optical scanners that can make a copy from an original.

(5) Audio-Input devices

Audio-input devices convert sounds into a form that can be processed by the system unit. By far the most widely used audio-input device is the microphone.

2. Output Devices

(1) Monitors

The most frequently used output device is the monitor. Also known as display screens, monitors present visual images of text and graphics. The output is often referred to as soft copy.

(2) Printers

You probably use a printer with some frequency to print homework assignments, photographs, and Web pages. Printers translate information that has been processed by the system unit and present the information on paper. Printer output is often called hard copy.

3. Combination Input and Output Devices

Many devices combine input and output capabilities. Sometimes this is done to save space. Other times it is done for very specialized applications. Common combination devices include fax machines, multifunctional devices, Internet telephones, and terminals.

NOTES

[1] full-sized, rigid keyboard 为原尺寸和刚性键盘, 就是常用的矩形键盘, 即传统键盘。flexible keyboard 是指可折叠、可卷起来的软键盘, 是一种小型键盘, 适合于装入袋中保存。

[2] mice 是 mouse 的复数形式。

KEYWORDS

hard copy	硬拷贝
soft copy	软拷贝
pointing	定位, 定点, 指向
pointer	指针
joystick	操作杆
touch screen	触摸屏
light pen	光笔
stylus	记录笔
mechanical mouse	机械鼠标

optical mouse	光学鼠标
cordless (wireless) mouse	无线鼠标
scanner	扫描仪
optical scanner	光学扫描仪
bar code reader	条形码阅读器
recognition device	识别设备
image capturing	图像捕获
audio-input	音频输入

EXERCISES

Fill in the blanks with appropriate words or phrases found behind this exercises.

1. Output on a screen is called as _____.
2. Output on a paper is called as _____.
3. Mouse, joysticks, touch screen, light pen, and stylus belong to _____.
4. Joystick is the most popular input device for _____.
5. When we use a light pen, it must be placed against the _____.
6. A stylus interacts with the computer through _____.
7. _____ convert scanned text and images into a form that the system unit can process.
8. _____ is one of the combination input and output devices.
9. By far the most widely used audio-input device is the _____.
10. _____ is commonly used in restaurants, ATMs, and information centers.

   a. Scanners

   b. computer games

   c. soft- copy

   d. Touch Screen

   e. hard-copy

   f. microphone

   g. pointing devices

   h. monitor

   i. Internet telephone

   j. hand-writing recognition program

CHAPTER 2 SYSTEM SOFTWARE

2.1 WINDOWS 10

Windows 10 is an upcoming operating system developed by Microsoft as part of the Windows NT family of operating systems. First presented in April 2014 at the Build Conference, it is scheduled to be released in mid-2015, and is currently in public beta testing through the Windows Insider program[1]. During its first year of availability, upgrades to Windows 10 will legally be offered at no charge for licensed consumer users of Windows 7 and Windows 8.1.

The goal of Windows 10 is to unify the Windows PC, Windows Phone, Windows Embedded and Xbox One product families around a common internal core[2]. These products will share a common, "universal" application architecture and Windows Store ecosystem that expands upon the Windows Runtime platform introduced by Windows 8[3]. Windows 10 will also introduce a new bundled web browser, Microsoft Edge, to replace Internet Explorer[4].

Figure 2-1 shows a Screenshot of Windows 10.

Figure 2-1   Screenshot of Windows 10

1. User interface and desktop

Windows 10's user interface is an evolution of Windows 8's, it changes its behavior depending on the type of device being used and available input methods. When a keyboard is attached, users are asked if they want to switch to a user interface mode that is optimized for mouse and keyboard, or stay within the touch-optimized mode. A new iteration of the Start menu is used, with an application list and the "All apps" button on the left side, and live tiles on the right. The menu can be resized, and expanded into a full-screen display, which is the default option in touch environments.

A new virtual desktop system known as Task View was added. Clicking the Task View button on the taskbar or swiping from the left side of the screen displays all open windows and allows users to switch between them, or switch between multiple workspaces[5]. Windows Store apps, which previously could only be used full-screen, can now be used in desktop windows or full-screen mode[6]. Program windows can now be snapped to quadrants of the screen by dragging them to the corner[7]. When a window is snapped to one side of the screen, the user is prompted to choose a second window to fill the unused side of the screen (called "Snap Assist")[8].

2. Features

A major aspect of Windows 10 is a focus on harmonizing user experiences between different classes of devices, along with addressing shortcomings in the Windows user interface that was first introduced in Windows 8[9].

The Windows Store app ecosystem has been revised into "Windows apps". They are made to run across multiple platforms and device classes, including smartphone, tablet, Xbox One, and other compatible Windows 10 devices. Windows apps share code across platforms, have responsive designs that adapt to the needs of the device and available inputs, can synchronize data between Windows 10 devices (including notifications, credentials, and allowing cross-platform multiplayer for games), and will be distributed through a unified Windows Store[10]. Developers can allow "cross-buys", where purchased licenses for an app apply to all of the user's compatible devices, rather than only the one they purchased on (i.e. a user purchasing an app on PC is also entitled to use the smartphone version at no extra cost)[11].

Windows 10 will also allow Web apps and desktop software (using either Win32 or .NET Framework) to be packaged for distribution on Windows Store[12]. Desktop software distributed through Windows Store will be packaged using the App-V system to allow sandboxing[13].

3. Windows 10 Editions

As in the past, we will offer different Windows editions that are tailored for various device families and uses. These different editions address specific needs of our various customers, from consumers to small businesses to the largest enterprises.

- Windows 10 Home is the consumer-focused desktop edition. It offers a familiar and personal experience for PCs, tablets and 2-in-1s.
- Windows 10 Mobile is designed to deliver the best user experience on smaller, mobile, touch-centric devices like smartphones and small tablets.
- Windows 10 Pro is a desktop edition for PCs, tablets and 2-in-1s. Building upon both the familiar and innovative features of Windows 10 Home, it has many extra features to meet the diverse needs of small businesses.
- Windows 10 Enterprise builds on Windows 10 Pro, adding advanced features designed to meet the demands of medium and large sized organizations.
- Windows 10 Education builds on Windows 10 Enterprise, and is designed to meet the needs of schools – staff, administrators, teachers and students.
- Windows 10 Mobile Enterprise is designed to deliver the best customer experience to business customers on smartphones and small tablets.

NOTES

[1] 长句，其中 First presented…为过去分词构成的时间状语从句，主句为 it is…，是并列句，主语 it 代表 Windows 10。Build 是微软主持的年会，旨在推进 Windows、Windows Phone、微软 Azure 和微软其他技术的使用。beta 是希腊 alphabet 的第二个字母，beta test 是软件开发中紧跟 alpha 第一阶段的第二阶段。在此阶段，软件功能基本实现，但还会有很多已知或未知的瑕疵（bugs）要发现和处理。Windows Insider program 是微软公司的程序，可使用户去访问微软软件的开发者。2014 年 9 月 30 日，该程序与 Windows 10 一起公布。2015 年 2 月 12 日用该程序为用户提供针对 Windows 10 的手机版预览。

[2] Windows Phone (WP)是微软为其智能手机开发的移动操作系统系列，取代了以前的 Windows Mobile 和 Zune 操作系统。由于微软新的商标策略和 Windows 10 的移动版在 2015 年面世，因此 Windows Phone 商标被淘汰。Windows Embedded 是微软用于嵌入式系统的操作系统系列。Xbox One 是微软在 2013 年 5 月 21 日公布并投放市场的家庭游戏控制台，它的前身是 Xbox 360。Xbox One 是 Xbox 系列中的第三个控制台。

[3] Windows Store 是微软发布和销售软件的平台，形象地说是一个商店。Windows Runtime (WinRT)是 2012 年 9 月在 Windows Server 2012 上首次提出的平台性质的应用体系结构。

[4] Microsoft Edge（代号 Spartan）是微软开发的万维网浏览器。官方首次公布是在 2015 年 1 月 21 日。作为预览的第一个公共版本是 2015 年 3 月 30 日公布的，它作为 Windows 10 PCs 智能手机和平板电脑的默认浏览器，取代了 IE。

[5] 长句。主语为现在分词短语 Clicking…or swiping…，谓语为 displays…and allows…。

[6] Windows Store app 是在 Windows 8 下的一种新型应用软件，是为触摸环境设计并优化的，比桌面应用软件更专业一些。

[7] Program windows 应为 Program for windows。

[8] Snap Assist 是 Windows 10 桌面侧边新加入的按钮，称为抓取助手，它可将多个不同桌面的应用展示在此，并和其他应用程序自由组合成多任务模式。

[9] 长句。主句是 A major aspect of Windows 10 is…，along with…介词短语作状语，其中 that 引导的定语从句修饰 interface。

[10] 长句。句子结构为 Windows apps share…，have…，can…，and will be…。其中 have…句中有 that 引导的定语从句。responsive design 响应式设计，实为 responsive Web design(RWD)，响应式页面设计，又称自适应页面设计，回应式页面设计，该设计可使网站在多种浏览设备（从桌面电脑显示器到移动电话或其他移动设备）上阅读和导航，同时减少缩放、平移和滚动。

[11] where 引导的是非限定性定语从句。cross-buys 是由 Sony 公司在 2012 年推出的，它可以让一个游戏在多个游戏网站上使用。

[12] Win32 是 WinAPI（正式应称为 Windows API）中的一个层次。WinAPI 是微软 Windows 操作系统中的应用程序界面的一套核心程序。.NET Framework 是微软开放的软件架构，主要运行在微软 Windows 上，它包括一个架构类库（Framework Class Library（FCL））的大型类库。

[13] App-V（Application Virtualization）微软的应用程序虚拟化，它是将应用程序转换为集中管理的服务。Sandbox，即沙盒、沙箱，是一种安全机制，为运行中的程序提供隔离环境，通常是为来源不可信、具有破坏力或无法判定程序意图的那些程序提供实验用。

KEYWORDS

codename	代号
availability	可用性，有效性，利用率，利用度
upgrade	升级，提高
licensed	特许的，许可的
architecture	体系结构，结构，层次结构，总体结构，结构格式
family	族，类，系列，种类
phased out	逐渐淘汰
console	控制台
embedded	嵌入式的
platform	平台
browser	浏览器
touch	触摸
button	按钮

taskbar	任务栏（条）
tile	平铺（并列）显示，铺切
resize	调整（重设）大小
framework	框架，构架，架构，体制组织
trackpad	跟踪垫（板），轨迹板
snap	抓取，快照，取图，快速移动
icon	图标，图符
default	默认，缺省
addressing	编址，寻址，定址
screenshot	屏幕快照
window	视窗
Snap Assist	抓取助手

EXERCISES

Multiple Choices

1. September 30, 2014 Microsoft released _____.
   a. Windows 7   b. Windows 10
   c. Build Conference   d. Windows Insider

2. Windows 10 has unified _____.
   a. Xbox One   b. Windows Phone
   c. Windows Embedded   d. Windows PC

3. Windows 10 introduced _____.
   a. IE   b. Microsoft Edge
   c. a new web browser   d. a bundled web browser

4. According to _____. Windows 10's user interface changes behavior of using computer.
   a. available input methods   b. available output methods
   c. the type of device being used d. the type of software being used

5. The Start menu in Windows 10 has_____.
   a. Live tiles   b. an application list
   c. "an app" button only   d. the "all apps" button

6. Task View button of Windows 10 _____.
   a. is known as virtual desktop system
   b. used for displaying all open windows
   c. used for switching between windows
   d. used for switching between multispaces

7. The Windows Store app ecosystem of Windows 10 can run _____.
   a. across multiple device classes   b. across multiple platforms
   c. smartphone   d. a platform only

8. Windows apps _____.

   a. can share code across platform

   b. have responsive design

   c. can synchronize data between Windows 10 devices

   d. will be distributed through a unified Windows Store

9. Data in Windows apps can be _____.

   a. notifications   b. credentials   c. input devices   d. games

10. App-V system _____.

   a. stands for Application Visualization   b. stands for Application Virtualization

   c. stands for Application Videocam   d. is used to package a Desktop software

11. Windows 10 Enterprise _____.

   a. builds on Windows 10 Mobile   b. used for small sized organizations

   c. builds on Windows 10 Pro   d. used for large sized organizations

12. Windows 10 Pro is _____.

   a. a desktop edition   b. used for PCs

   c. used for tablets   d. used for small businesses

2.2 UNIX AND LINUX

1. UNIX

UNIX is an operating system originally developed by Dennis Ritchie and Ken Thompson at AT &T Bell Laboratories that allows a computer to handle multiple users and programs simultaneously. Since its development in the early 1970s, UNIX has been enhanced by many individuals and particularly by computer scientists at the University of California, Berkeley (known as Berkeley Software Distribution UNIX, or BSD UNIX)[1]. This operating system is available on a wide variety of computer systems, ranging from personal computers to mainframes, and is available in other related forms[2]. AIX is an implementation that runs on IBM workstations, A/UX is a graphical version that runs on Macintosh computers; Solaris runs on Intel microprocessors[3].

Features:

① The UNIX system can support multi-users and multi-tasks

② The UNIX System Kernel

The kernel is the heart of the UNIX operating system, responsible for controlling the computer's resources and scheduling user jobs so that each one gets its fair share of the resources. Programs interact with the kernel through special functions with well-known names, called system calls.

③ The shell

The shell is a command interpreter that acts as an interface between users and the operating system. When you enter a command at a terminal, the shell interprets the command and calls the program you want.

④ Device-Independent Input and Output

Devices (such as a printer or terminal) and disk files all appear as files to UNIX programs. When you give the UNIX operating system a command, you can instruct it to send the output to any one of several devices or files. This diversion is called output redirection.

2. Linux

Linux is a UNIX-like and mostly POSIX-compliant computer operating system assembled under the model of free and open-source software development and distribution[4]. The defining component of Linux is the Linux kernel, an operating system kernel first released on 5 October 1991 by Linus Torvalds. The Free Software Foundation uses the name GNU/Linux to describe the operating system, which has led to some controversy[5].

Linux was originally developed as a free operating system for Intel x86–based personal computers, but has since been ported to more computer hardware platforms than any other operating system[6]. It is the leading operating system on servers and other big iron systems such as mainframe computers and supercomputers, but is used on only around 1.5% of desktop computers[7]. Linux also runs on embedded systems, which are devices whose operating system is typically built into the firmware and is highly tailored to the system. Android, the most widely used operating system for tablets and smartphones, is built on top of the Linux kernel[8].

(1) Hardware support

Linux kernel is a widely ported operating system kernel; it runs on a highly diverse range of computer architectures, including the hand-held ARM-based iPAQ and the IBM mainframes System z9 or System z10—covering devices ranging from mobile phones to supercomputers, as showing as Figure 2-2[9].

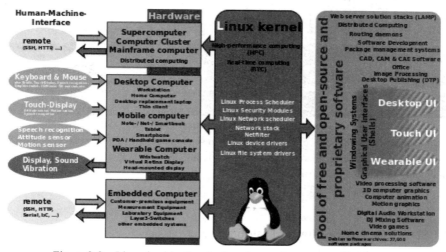

Figure 2-2　Linux is ubiquitously found on various types of hardware

(2) Uses

Beside the Linux distributions designed for general-purpose use on desktops and servers, distributions may be specialized for different purposes including: computer architecture support, embedded systems, stability, security, localization to a specific region or language, targeting of specific user groups, support for real-time applications, or commitment to a given desktop environment[10]. As of 2015, over four hundred Linux distributions are actively developed, with about a dozen distributions being most popular for general-purpose use.

(3) Desktop

The popularity of Linux on standard desktop computers and laptops has been increasing over the years. Currently, most distributions include a graphical user environment, with the two most popular environments being GNOME (which can utilize additional shells such as the default GNOME Shell and Ubuntu Unity) and the KDE Plasma Desktop[11].

No single official Linux desktop exists: rather desktop environments and Linux distributions select components from a pool of free and open-source software with which they construct a GUI implementing some more or less strict design guide[12]. GNOME, for example, has its human interface guidelines as a design guide, which gives the human–machine interface an important role, not just when doing the graphical design, but also when considering people with disabilities, and even when focusing on security[13].

Figure 2-3 shows the Visible software components of the Linux desktop stack.

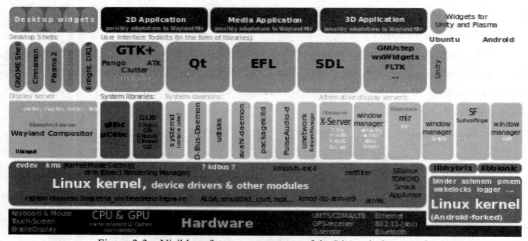

Figure 2-3　Visible software components of the Linux desktop stack

Visible software components of the Linux desktop stack include the display server, widget engines, and some of the more widespread widget toolkits. There are also components not directly visible to end users, including D-Bus and PulseAudio[14].

NOTES

[1] BSD 即 Berkeley Software Distribution，原意为伯克利软件发行中心，可直译为伯克利版软件或伯克利 UNIX。

[2] 由 and 连接的并列句，前一句中 ranging…为分词短语作状语。

[3] 此处介绍了 UNIX 的几个变种，其中 Macintosh 机器上的 A/UX 是苹果公司开发的 UNIX，Solaris 是运行在 SUN 公司工作站上的操作系统，其 CPU 使用的是 Intel 微处理器。

[4] POSIX，Portable Operating System UNIX，可移植的 UNIX 系统。

[5] Free Software Foundation (FSF)，即自由软件基金会，是 1985 年成立的非营利组织，致力于自由软件的推广。GNU(Gnu's Not UNIX)是一个自由软件组织。Which 引导的是非限定性定语从句。

[6] 并列句，后一句省略了主语 Linux。Intel x86 是 Intel 公司生产的微处理器芯片，x 代表 802，803，804…。

[7] big iron 在计算机领域中是指大型、昂贵、非常快速的计算机，常用于描述像格雷（Grays）那样的超级计算机，也用来指大型商业 IBM 计算机等。

[8] Android 是用于智能电话和平板电脑等移动设备的操作系统。

[9] ARM-based 是 Arm 公司的微处理器和微控制器集成电路，是基于各种 32 位的 ARM 处理器芯片。iPAQ 是指 2000 年 4 月由康柏（Compaq）公司首次公布的袖珍电脑和个人数字助理（PDA），此名称是借用康柏公司早期 iPAQ 台式个人计算机名字。

[10] 长句。Beside 此处为介词，由它引出的介词短语作状语，主句为 distributions may be…。Linux distributions 是 Linux 发行版，是为一般用户预先集成好的 Linux 操作系统及各种应用软件，通常包括桌面环境办公套件、媒体播放器、数据库等应用软件。

[11] GNOME 是一个桌面环境，它全部由自由和开源软件组成，其目标操作系统是 Linux，但也支持其他大多数 BSD 衍生软件。GNOME Shell 是 GNOME 桌面环境的正式用户界面。Ubuntu Unity 中，Unity 是由 Canonical 有限公司为其 Ubuntu 操作系统开发的 GNOME 桌面环境中的图形外壳。KDE 是一个国际自由软件团体，它是制作在 Linux、FreeBSD、Solaris、Microsoft Windows 和 OS X 系统上的、集成化、跨平台的应用软件系列。这种称为 Plasma Desktop 的，是一个桌面环境。

[12] with which 引导的定语从句，修饰 components。

[13] 长句，which 引导的是非限定性定语从句。

[14] D-Bus 实质上是一个适用于桌面应用进程间的通信机制，即所谓的 IPC（Inter-Process Communication）机制。PulseAudio（以前叫 Polypaudio）是一个跨平台的、可通过网络工作的声音服务，其一般使用于 Linux 和 FreeBSD 操作系统。

KEYWORDS

UNIX	一种计算机操作系统
workstation	工作站
multi-user	多用户
multi-task	多任务
job	作业
kernel	内核

interface	接口，界面
system call	系统调用
interpreter	解释程序
disk file	磁盘文件
redirection	重定向
device-independent	设备无关的
shell	外壳（程序）、命令解释程序
Linux	以 Linux 命名的操作系统
Gnu's Not UNIX (GNU)	一个自由软件组织
embedded system	嵌入式系统
firmware	固件
tailor	设计，裁剪，加工，处理，编（特）制
port	移植，端口
localization	局部化，地方化，本地化
security	安全，安心，防护，保障
widget	小装置（窗口），窗口软件设计工具，界面构造（组件），图形设备，数据类型
GUI(Graphic User Interface)	图形用户界面
human-machine interface	人机界面

EXERCISES

Fill in the blanks with appropriate words or phrases found behind this exercises.

1. UNIX can be used on _____.
2. UNIX System can support _____ users and _____ tasks.
3. The UNIX allows each user to run _____ job at a time.
4. _____ are the special functions that interact programs with the kernel of UNIX.
5. The shell is an interface between _____.
6. You can instruct _____ to send the output to any one of several devices or files.
7. Linux is a _____ operating system.
8. FSF uses the name _____ to describe Linux.
9. _____ is the leading operating system on servers.
10. _____ has a built-in operating system.
11. _____ is built on top of the Linux kernel.
12. _____ is a graphical user environment.

    a. UNIX-like

    b. UNIX

    c. embedded system

    d. PCs, minicomputers, mainframes

e. users, and the operating system

f. GNOME

g. more than one

h. Linux

i. Android

j. GNU/Linux

k. system calls

l. many,many

2.3 ANDROID

Android is an operating system for touchscreen mobile devices such as smartphones and tablet computers. It is developed by the Open Handset Alliance—a consortium of hardware, software, and telecommunication companies led by Google.[1]

Android consists of a kernel based on the Linux kernel, with middleware, libraries and APIs written in C and application software running on an application framework which includes Java-compatible libraries based on Apache Harmony.[2]

As of July 2013, the Google Play store has had over one million Android applications ("apps") published, and over 50 billion applications downloaded. A developer survey conducted in April—May 2013 found that 71% of mobile developers develop for Android[3]. At Google I/O 2014, the company revealed that there were over one billion active monthly Android users, up from 538 million in June 2013. As of 2015, Android has the largest installed base of all general-purpose operating systems[4].

Android's source code is released by Google under open source licenses, although most Android devices ultimately ship with a combination of open source and proprietary software, including proprietary software developed and licensed by Google.

Android is popular with technology companies which require a ready-made, low-cost and customizable operating system for high-tech devices. Android's open nature has encouraged a large community of developers and enthusiasts to use the open-source code as a foundation for community-driven projects, which add new features for advanced users or bring Android to devices which were officially released running other operating systems[5]. The operating system's success has made it a target for patent litigation as part of the so-called "smartphone wars" between technology companies.

Current features and specifications in handset layouts include

(1) Platform

The platform is adaptable to larger, VGA, 2D graphics library, 3D graphics library based on OpenGL ES 2.0 specifications, and traditional smartphone layouts [6].

(2) Storage

SQLite, a lightweight relational database, is used for data storage purposes.

(3) Connectivity

Android supports connectivity technologies including GSM/EDGE, IDEN, CDMA, EV-DO, UMTS, Bluetooth, Wi-Fi, and etc [7].

(4) Messaging

SMS and MMS are available forms of messaging, including threaded text messaging and now C2DM is also a part of Android Push Messaging service [8].

(5) Web browser

The web browser available in Android is based on the open-source WebKit layout engine, coupled with Chrome's V8 JavaScript engine [9].

(6) Java support

While most Android applications are written in Java, there is no Java Virtual Machine in the platform and Java byte code is not executed [10].

(7) Media support

Android supports the following audio/video/still media formats: MP3, MIDI, WAV, JPEG, PNG, GIF, BMP and so forth [11].

(8) Streaming media support

RTP/RTSP streaming, HTML progressive download, Adobe Flash Streaming (RTMP) and HTTP Dynamic Streaming are supported [12].

(9) Additional hardware support

Android can use televisions (Android TV), car (Android Auto), wrist watches (Android Wear), still/digital cameras, game consoles, touchscreens, GPS, etc.

(10) Video calling

Android does not support native video calling, but some handsets have a customized version of the operating system that supports it.

There are another supports in handset layouts, they are multiple language, multitasking, multi-touch, bluetooth, voice based features, and screen capture.

NOTES

[1] Open Handset Alliance，开放手机联盟，是美国 Google 公司于 2007 年 11 月 5 日宣布组建的一个全球性的联盟组织。

[2] 本段就是一句话，句中大部分是 with…引导的状语，其中还有 running…现在分词短语和 which 引导的定语从句。Linux 内容见 2.4 节。Apache Harmony 是 Apache 软件基金会主导的开放源代码专题，是自由 JAVA 实现计划（Free Java implementations）的一部分。有关 Java 详见 6.2 节。

[3] conducted…过去分词短语作定语，修饰 survey。found 为过去时，作谓语。that 引导的是宾语从句。

[4] Google I/O 2014，是谷歌在美国加州旧金山召开的软件开发者年会，其主旨是在使用谷歌和开放万维网技术，如 Android、Chrome、Chrome OS、Google APIs、Google Web Kit、App Engine 等方面，采用高科技构建万维网、移动和企业应用等问题。

[5] which add…or bring…是非限定性定语从句，后一个 which 引导的是定语从句，修饰 devices。

[6] VGA（Video Graphics Array）即视频图形阵列，是 IBM 在 1987 年随 PS/2 机推出的使用模拟信号的一种视频传输标准。这个标准对于现今的个人计算机市场已经十分过时。即便如此，VGA 仍然是最多制造商支持的一个标准。

OpenGL ES (OpenGL for Embedded Systems) 是 OpenGL 三维图形 API 的子集，针对手机、PDA 和游戏主机等嵌入式设备设计。

[7] GSM 是全球移动通信系统（Global System for Mobile Communications）的缩写，俗称"全球通"。EDGE 是英文 Enhanced Data Rate for GSM Evolution 的缩写，即增强型数据速率 GSM 演进技术。EDGE 是一种从 GSM 到 3G 的过渡技术。

IDEN（集成数字增强型网络）系统是摩托罗拉公司研制和生产的一种数字集群移动通信系统。CDMA 是码分多址的英文缩写（Code Division Multiple Access），它是在数字技术的分支——扩频通信技术上发展起来的一种崭新而成熟的无线通信技术。

EV-DO 是英文 Evolution-Data Optimized 或者 Evolution-Data Only 的缩写，有时也写作 EVDO 或者 EV。其中 CDMA2000 1xEV-DO 是一种可以满足移动高速数据业务的技术。UMTS（Universal Mobile Telecommunications System）即通用移动通信系统。

Bluetooth、Wi-Fi 见 3.1.2 节。

[8] SMS 是一种存储和转发服务。也就是说，短消息并不是直接从发送人发送到接收人，而是通过 SMS 中心进行转发。MMS 为 Multimedia Messaging Service 的缩写，中文译为多媒体短信服务，即彩信。C2DM 是谷歌的一种服务，它允许开发者从服务器向 Android 应用或 Chrome apps 发送数据。

[9] WebKit 是在 Web 浏览器中用来绘制网页的排版引擎软件。WebKit 目前是 Apple Safari 及 Google Chrome 等浏览器的主要引擎。Chrome's V8 JavaScript engine 是谷歌为万维网浏览器 Google Chrome 开发的一个开源 JavaScript 引擎。

[10] Java Virtual Machine（JVM）是一个抽象计算机，它由三个概念构成：规范、实现和范例。

[11] MIDI 是 Musical Instrument Digital Interface 的缩写，意思是音乐设备数字接口。这种接口技术的作用就是使电子乐器与电子乐器，电子乐器与计算机之间通过一种通用的通信协议进行通信，这种协议自然就是 MIDI 协议了。WAV 为微软公司开发的一种声音文件格式，它符合 RIFF（Resource Interchange File Format）文件规范，用于保存 Windows 平台的音频信息资源，被 Windows 平台及其应用程序所广泛支持。JPEG 是联合图像专家组（Joint Photographic Experts Group）的首字母缩写，是一种广泛适用的压缩图像标准格式。PNG 即图像文件存储格式，其目的是试图替代 GIF 和 TIFF 文件格式，同时增加一些 GIF 文件格式所不具备的特性。GIF（Graphics Interchange Format）是图像互换格式。BMP 是英文 Bitmap（位图）的简写，它是 Windows 操作系统中的标准图像文件格

式，能够被多种 Windows 应用程序所支持。

[12] RTP（Real-time Transport Protocol，实时传送协议）是一个网络传输协议。RTSP（Real Time Streaming Protocol，实时流协议）是应用级协议，控制实时数据的发送。HTML（HyperText Markup Language，超文本标记语言）是一种专门用于创建 Web 超文本文档的编程语言。HTTP Dynamic Streaming（动态流）已逐渐成为因特网上主流的媒体形式，占据着不可替代的位置。

KEYWORDS

middleware	中间件
API (Application Program Interface)	应用程序接口
framework	框架，构架，架构，体制组织
smartphone	智能手机，智能电话
lightweight relational database	小型关系型数据库
streaming media	流媒体
threaded text messaging	线程文本消息
source code	源代码
proprietary software	专有软件，专利软件
alliance	联合，联盟
consortium	会，社，联盟，合作，财团，联营企业，国际性企业
telecommunication	电信，远程通信
layout	布局，布置，格式方案，规划，设计，草图，轮廓

EXERCISES

Fill in the blanks with using appropriate words or terms found behind this exercises.

1. Android is developed by _____.
2. Android has been deployed with _____.
3. SQLite is a _____.
4. Android's connectivity includes _____ and so forth.
5. Android's Cloud To Device Messaging can be abbreviated to _____.
6. The Web browser available in Android is based on _____.
7. Android has a customized version of the operating system that supports _____.
8. Function of wrist watches can be performed with _____ software.
9. Android has the largest installed base of all _____.
10. Most Android devices ship with a combination of _____ software.
11. The Open Handset Alliance is a _____ of hardware, software, and telecommunication companies.

12. The open-source code used for Android is as a _____ for community-driven projects.
13. Android can be brought to devices running _____.
14. "Smartphone wars" are caused by Android's _____.

    a. video calling

    b. the open-source WebKit layout engine

    c. other operating system

    d. open source and proprietary

    e. Android Wear

    f. success

    g. foundation

    h. general-purpose operating system

    i. C2DM

    j. GSM, CDMA, Wi-Fi, Bluetooth

    k. lightweight relational database

    l. consortium

    m. the Open Handset Alliance

    n. middleware, libraries and APSs

CHAPTER 3   COMPUTER NETWORK

3.1   LOCAL AREA NETWORKS (LANs)

3.1.1   ETHERNET

A local area network (LAN) is a computer network that interconnects computers within a limited area such as a home, school, computer laboratory, or office building, using network media[1]. The defining characteristics of LANs, in contrast to wide area networks (WANs), include their smaller geographic area, and non-inclusion of leased telecommunication lines.

ARCNET, Token Ring and other technology standards have been used in the past, but Ethernet over twisted pair cabling and Wi-Fi are the two most common technologies currently used to build LANs [2].

Network topology describes the layout of interconnections between devices and network segments. At the Data Link Layer and Physical Layer, a wide variety of LAN topologies have been used, including ring, bus, mesh, and star, but the most common LAN topology in use today is switched Ethernet[3]. At the higher layers, the Internet Protocol (TCP/IP) has become the standard, replacing NetBEUI, IPX/SPX, AppleTalk and others [4].

Simple LANs generally consist of one or more switches. A switch can be connected to a router, cable modem, or ADSL modem for Internet access. Complex LANs are characterized by their use of redundant links with switches using the spanning tree protocol to prevent loops, their ability to manage differing traffic types via quality of service (QoS), and to segregate traffic with VLANs[5]. A LAN can include a wide variety of network devices such as switches, firewalls, routers, load balancers, and sensors.

LANs can maintain connections with other LANs via leased lines, leased services, or the Internet using virtual private network (VPN) technologies.

1. Network switch

A network switch (also called switching hub, bridging hub, officially MAC bridge) is a computer networking device that connects devices together on a computer network (Figure 3-1), by using packet switching to receive, process and forward data to the destination device[6]. Unlike less advanced network hubs, a network switch forwards data only to one or multiple devices that need to receive it, rather than broadcasting the same data to each of its ports.

Figure 3-1　Avaya ERS 2550T-PWR, a 50-port Ethernet switch

2. 10 Gigabit Ethernet

10 Gigabit Ethernet (10GE, 10GbE, or 10 GigE) is a group of computer networking technologies for transmitting Ethernet frames at a rate of 10 gigabits per second (10×10^9 or 10 billion bits per second). It was first defined by the IEEE 802.3ae-2002 standard[7]. Unlike previous Ethernet standards, 10 Gigabit Ethernet defines only full duplex point-to-point links which are generally connected by network switches; shared-medium CSMA/CD operation has not been carried over from the previous generations Ethernet standards. Half duplex operation and hubs do not exist in 10GbE.

3. Terabit Ethernet

Facebook and Google, among other companies, have expressed a need for TbE[8]. Some think that a speed of 400 Gbit/s is a more practical goal than 1Tbit/s (1000Gbit/s). In 2011, researchers predicted Terabit Ethernet (1Tbit/s) in 2015, and 100 Terabit Ethernet (100Tbit/s) by 2020.

NOTES

[1] network media，此处指传输媒体，或传输介质，如双绞线、光纤和无线传播等。

[2] 长句。ARCNET 原文是 Attached Resource Computer NETwork，是一种局域网通信协议，在 20 世纪 80 年代微型计算机网络中较为常用。随着以太网的出现和快速普及，ARCNET 已退出历史舞台。Token Ring 即令牌环，是环形局域网采取的工作方式，现已不多用。Wi-Fi，见 3.1.2 节。

[3] Data Link Layer and Physical Layer 分别是国际标准化组织（ISO）制定的开放系统互连协议（OSI）七层中的第二层和第一层，即数据链路层和物理层。

[4] TCP/IP（Transmission Control Protocol/Internet Protocol）即传输控制协议/互联网协议，是因特网最重要的协议。NetBEUI 和 IPX/SPX 是早期局域网 Novell 使用的协议。AppleTalk 是苹果公司为其 Macintosh 计算机开发的网络协议簇，有很强的网络功能。

[5] 长句。主句为 Complex LANs are characterized by…，by 后面为两个并列的方式状语。spanning tree protocol，生成树协议，即生成树算法。QoS（Quality of Service）即服务质量，在计算机网络中，是流量工程中的术语，是一种控制机制。VLAN（Virtual LAN）即虚拟局域网。

[6] 长句。主句是 A network switch is…，其中 that 引导的定语从句，修饰 device，

by using…分词短语作状语。

[7] IEEE 802.3 是电气电子工程师协会（Institute of Electrical and Electronic Engineers）IEEE 802 课题为局域网制定的标准。这一标准采用 CSMA/CD 访问方法，是以太网的正式标准。

[8] Facebook 即脸谱网站，详见 5.1.2 节。

KEYWORDS

LAN (Local Area Network)	局域网
WAN (Wide Area Network)	广域网
telecommunication	电信，远程通信
twisted pair	双绞线（对）
Wi-Fi (Wireless Fidelity)	无线保真
network topology	网络拓扑(结构)
ring	环
bus	总线
mesh	网格
star	星形
Switched Ethernet	交换式以太网
switch	交换机
router	路由器
cable modem	电缆调制解调器
ADSL(Asymmetric Digital Subscriber Line)	非对称数字用户线
link	链接，链路
redundant links	冗余链路
spanning tree	生成树
loop	环，环路
traffic	通信数据量，业务量
QoS(Quality of Service)	质量保证
firewall	防火墙
load balance	负载均衡
sensor	传感器
leased	租用
VPN(Virtual Private Network)	虚拟专用网
hub	集线器
bridge	桥，网桥
MAC(Medium Access Control)	媒体访问（接入，存取）控制
packet switching	分组交换
forward	转发

broadcasting	广播
port	端口，移植
Gigabit	吉比特，吉位
Tb(Terabit)	太位
frame	帧
full-duplex	全双工，双向同时通信
half-duplex	半双工通信，双向交替通信
point-to-point	点对点
shared-medium	共享媒体

EXERCISES

Fill in the blanks with appropriate words or phrases found behind this exercises.

1. A LAN is a computer network that connects _____.
2. Network media we used here include _____.
3. The two most common technologies currently used to build LANs are _____.
4. _____ describes the layout of interconnections between devices and network segments.
5. _____ is the most common LAN topology in use today.
6. Spanning tree topology is used for _____.
7. VLAN stands for _____.
8. Network devices consisting a LAN can be _____.
9. Switching hub, bridging hub, and MAC bridge are the same as _____.
10. Network switch uses _____ to transmit data.
11. Official Ethernet standard uses _____ access method.
12. _____ is the term used for traffic engineering in computer network.

    a. switches, firewalls, routers, load balancers, and sensors
    b. QoS
    c. twisted pair, optical fiber, and wireless
    d. Network topology
    e. packet switching
    f. complex LANs
    g. computers within a limited area
    h. network switch
    i. CSMA/CD
    j. Virtual LANs
    k. Switched Ethernet
    l. twisted pair cabling and Wi-Fi

3.1.2 WI-FI AND BLUETOOTH

1. Wi-Fi

Wi-Fi (or, incorrectly but commonly, WiFi) is a local area wireless technology that allows an electronic device to participate in computer networking using 2.4GHz UHF and 5GHz SHF ISM radio bands[1].

The Wi-Fi Alliance defines Wi-Fi as any "wireless local area network" (WLAN) product based on the Institute of Electrical and Electronics Engineers' (IEEE) 802.11 standards. However, the term "Wi-Fi" is used in general English as a synonym for "WLAN" since most modern WLANs are based on these standards. "Wi-Fi" is a trademark of the Wi-Fi Alliance. The "Wi-Fi CERTIFIED" trademark can only be used by Wi-Fi products that successfully complete Wi-Fi Alliance interoperability certification testing[2].

Many devices can use Wi-Fi, e.g. personal computers, video-game consoles, smartphones, digital cameras, tablet computers and digital audio players. These can connect to a network resource such as the Internet via a wireless network access point. Such an access point (or hotspot) has a range of about 20 meters (66 feet) indoors and a greater range outdoors. Hotspot coverage can comprise an area as small as a single room with walls that block radio waves, or as large as many square kilometers achieved by using multiple overlapping access points[3]. Figure 3-2 depicts the communication course between a notebook and a printer through an access point.

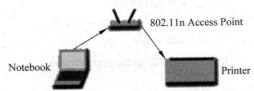

Figure 3-2   Depiction of a device sending information wirelessly to another device,
both connected to the local network, in order to print a document.

Wi-Fi can be less secure than wired connections, such as Ethernet, because an intruder does not need a physical connection. Web pages that use SSL are secure but unencrypted internet access can easily be detected by intruders[4]. Because of this, Wi-Fi has adopted various encryption technologies. The early encryption WEP proved easy to break[5]. Higher quality protocols (WPA, WPA2) were added later[6]. An optional feature added in 2007, called Wi-Fi Protected Setup (WPS), had a serious flaw that allowed an attacker to recover the router's password[7]. The Wi-Fi Alliance has since updated its test plan and certification program to ensure all newly certified devices resist attacks.

2. Bluetooth

Bluetooth is a wireless technology standard for exchanging data over short distances (using short-wavelength UHF radio waves in the ISM band from 2.4 to 2.485GHz) from fixed and mobile devices, and building personal area networks (PANs) [8]. Invented by telecom vendor Ericsson in 1994, it was originally conceived as a wireless alternative to RS-232 data cables. It can connect several devices, overcoming problems of synchronization.

Bluetooth is managed by the Bluetooth Special Interest Group (SIG), which has more than 25,000 member companies in the areas of telecommunication, computing, networking, and consumer electronics[9]. The IEEE standardized Bluetooth as IEEE 802.15.1, but no longer maintains the standard. The Bluetooth SIG oversees development of the specification, manages the qualification program, and protects the trademarks. A manufacturer must make a device to meet Bluetooth SIG standards, and market it as a Bluetooth device.

Bluetooth is a standard wire-replacement communications protocol primarily designed for low-power consumption, with a short range based on low-cost transceiver microchips in each device. Because the devices use a radio (broadcast) communications system, they do not have to be in visual line of sight of each other, however a quasi optical wireless path must be viable[10]. Range is power-class-dependent, but effective ranges vary in practice; see Figure 3-3.

Class	Max. permitted power (mW)	(dBm)	Typ. range (m)
1	100	20	~100
2	2.5	4	~10
3	1	0	~1

Figure 3-3   Class of Bluetooth

3. Bluetooth VS Wi-Fi

Bluetooth and Wi-Fi have some similar applications: setting up networks, printing, or transferring files. Wi-Fi is intended as a replacement for high speed cabling for general local area network access in work areas. This category of applications is sometimes called wireless local area networks (WLAN). Bluetooth was intended for portable equipment and its applications. The category of applications is outlined as the wireless personal area network (WPAN). Bluetooth is a replacement for cabling in a variety of personally carried applications in any setting and also works for fixed location applications such as smart energy functionality in the home (thermostats, etc.) [11].

Wi-Fi and Bluetooth are to some extent complementary in their applications and usage. Wi-Fi is usually access point-centered, with an asymmetrical client-server connection with all traffic routed through the access point, while Bluetooth is usually symmetrical, between two Bluetooth devices[12]. Bluetooth serves well in simple applications where two devices need to connect with minimal configuration like a button press, as in headsets and remote controls, while Wi-Fi suits better in applications where some degree of client configuration is possible and high speeds are required, especially for network access through an access node[13].

NOTES

[1] that 引导的定语从句，修饰 technology，using…分词短语作状语。UHF（Ultra High Frequency）、SHF（Super-High Frequency）都是超高频，但频带范围不同，UHF 为 300MHz～3GHz，SHF 为 3～30GHz。ISM（Industrial Science Medicine）频段是由 ITU-R（ITU Radio communication Sector，国际电信联盟无线电通信组）定义的。

[2] that 引导的定语从句，修饰 products。

[3] or 连接的并列长句，两句中分别有 that 引导的定语从句和过去分词 achieved 短语构成的定语从句。

[4] SSL（Secure Sockets Layer），安全套接层，计算机网上提供通信安全的加密协议。

[5] WEP（Wired Equivalent Privacy），与有线网等效的保密协议，是早期（1999 年 9 月批准的）为 IEEE 802.11 设计的无线网络安全算法。

[6] WPA（Wi-Fi Protected Access）和 WPA2 是 Wi-Fi 联盟开发的，用于保护无线计算机网络安全的安全协议和认证程序。

[7] WPS（Wi-Fi Protected Setup），是一个网络安全标准，目的是使用户较容易地保护无线家庭网络。

[8] 句中 for exchanging…，and building…是目的状语从句。

[9] Bluetooth Special Interest Group（SIG）即蓝牙专门兴趣组，是非营利组织，主要任务是监督蓝牙标准的发展，蓝牙技术和商标的授权等情况。

[10] line of sight 即视距（传播），是电磁辐射或声波传播的一个特性，即直线传播特性。

[11] 由 and 连接的并列句。

[12] 并列长句。第一个 with 引导的是方式状语从句；第二个 with 引导的是定语，修饰 connection。

[13] 并列长句，两个句子中都有 where 引导的定语从句。

KEYWORDS

Wi-Fi (Wireless Fidelity)	无线保真
UHF(Ultra High Frequency)	特高频
SHF(Super-High Frequency)	超高频
ISM(Industrial Science Medicine)	工业、科学和医用频段
WLAN(Wireless Local Area Network)	虚拟局域网
certified	已证明无误的，合格的
video-game console	电视游戏控制台
smartphone	智能手机

digital camera	数码相机
tablet computer	平板电脑
digital audio player	数字音频播放器
access point	接入点，存取点，访问点
hotspot	热点
radio wave	无线电波
overlap	重叠，部分重叠
intruder	侵入者，干扰者
SSL(Secure Sockets Layer)	安全套接层
detect	检测，探测，检波
encryption	加密
unencryption	解密
flaw	缺陷，瑕疵，裂缝，故障
router	路由器
password	密码，口令
certification	证明（书），认可
Bluetooth	蓝牙
PAN(Personal Area Network)	个人区域网
telecom	电信，远程通信
vendor	自动售货机，卖主，厂商，供应商
qualification	资格，限定，合格性
protect	保护，防御
consumption	消费，消耗
transceiver	收发器，收发信机
microchip	微[芯]片
line of sight	视线，视距
thermostat	自动调温器
complementary	互补的，附加的
optical wireless	光无线电
symmetrical	对称的
asymmetrical	非对称的
client-server	客户-服务器
configuration	配置，排列，结构，外形，形状
suit	一套，一组，适合于
access node	访问节点，接入节点

EXERCISES
Multiple choices
1. Wi-Fi _____.
    a. is a wireless technology of LAN
    b. uses 2.4GHz UHF bands
    c. can be written to WiFi
    d. complies IEEE 802.11 standards
2. A wireless network access point _____.
    a. can cover about 20 meters indoors
    b. is known as hotspot
    c. can cover about several square kilometers
    d. can cover about 100 meters indoors
3. Wi-Fi network _____.
    a. is secure enough
    b. is less secure
    c. is secure with using SSL technology
    d. is less secure with unencrypted internet access
4. Devices that can use Wi-Fi include _____.
    a. tablet computer
    b. video-game consoles
    c. digital cameras
    d. smartphones
5. Higher quality protocol of security is _____.
    a. WEP
    b. WPA
    c. WPA2
    d. WPS
6. Bluetooth is used for exchanging data _____.
    a. over short distances
    b. over longer distances
    c. between mobile devices only
    d. between fixed and mobile devices
7. Members of SIG are in the area of _____.
    a. computing
    b. telecommunication
    c. networking
    d. consumer electronics
8. Tasks of SIG are _____.
    a. managing the qualification program
    b. protecting the trademarks
    c. keeping watch on the development of the Bluetooth standard
    d. maintaining the IEEE standard
9. Devices used in Bluetooth are _____.
    a. higher power consumption
    b. low power consumption
    c. low cost devices
    d. with a broadcast communication system
10. The effective range of Bluetooth communication is _____ meters.
    a. one
    b. ten
    c. one hundred
    d. one thousand
11. Wi-Fi and Bluetooth have some similar applications _____.
    a. transferring file
    b. asymmetrical client/server connection
    c. setting up network
    d. printing
12. Bluetooth and Wi-Fi are to some extent complementary in _____.

a. Wi-Fi is access point-centered

b. Bluetooth is two devices accessed

c. Wi-Fi suits for network access via an access node

d. Bluetooth suits for simple applications

3.2 THE INTERNET

The Internet is the largest and most well-known computer network in the world. It is technically a network of networks since an individual user connects to a network set up by their access provider or Internet service provider (ISP), which in turn is connected to a larger network, which may be connected to an even larger network [1]. Altogether, this network of networks is referred as the Internet. Since all the networks on the Internet are interconnected, any computer with Internet access can communicate with any other computer on the Internet, regardless of the ISP used.

1. The Domain Name System (DNS)

The Domain Name System (DNS) was created to centralize the task of making changes to the network name to address assignments and to automate the task of performing the translation functions[2]. In the early days of the Internet, a central location (SRI NIC at Stanford Research Institute in Melo Park, California) was responsible for maintaining a HOSTS file that contained the name of every host on the Internet along with its address [3]. Administrators had to communicate changes to SRI NIC, and these changes were incorporated into the file periodically. Of course, this meant that the file then had to be distributed to every single host so that it could have the updated version [4].

DNS uses a hierarchical distributed architecture that is spread across many computers throughout the Internet. A root server holds information about top level domains (such as .COM, .EDU, and .GOV), and each domain throughout the Internet has a domain name server that is responsible for the computer names and addresses used in that domain [5]. Client computers query DNS servers when they need to get the address for a hostname. If the local DNS knows the address, it returns it to the client computer. If it does not, it sends the query up the chain of DNS servers until a DNS server is found that can resolve the name, provided that it is indeed a valid name [6].

The topmost entry in the DNS hierarchy is called the root domain and its represented by the period character (.). Underneath this root domain are the top-level directories that fall into two groups: geographical and organizational. Geographical domains are used to specify specific countries. For example, .au for Australia and .cn for China. Under each of the geographical domains, you can find organizational domains.

Organizational domains you might be familiar with include the following [7]:

(1) com   Used for commercial organizations.
(2) edu   Used for educational institutions.
(3) gov   Used for U.S. Government entities.
(4) mil   Used for U.S. military organizations.
(5) Int   International organizations.
(6) net   Used for Network organizations such as Internet service providers.
(7) org   Used for nonprofit organizations.
(8) arpa   Used for inverse address lookups.

The structure of the Domain Name System is similar to an inverted tree. In Figure 3-4 you can see that at the top is the root domain with the com through cn domains underneath [8]. Under the com domain are individual business organizations that each their own domain. Under any particular domain, there can be sub-domains.

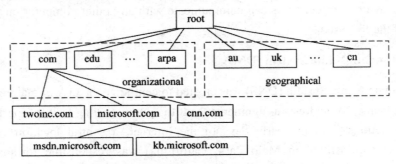

Figure 3-4   Domain Name System is a distributed, hierarchical structure

2. Connecting To The Internet

(1) Dial-Up Connection

Dial-Up Connection usually work over regular telephone lines. To connect to the Internet your modem (or other appropriate interface devices) dials up and connects to a modem attached to your ISP's computer [9]. While you are connected, your PC is assigned a temporary IP address for the current session [10]. At the end of each Internet session, you disconnect from your ISP's computer to allow another user to connect in your place[11]. Standard dial-up Internet service uses a conventional dial-up modem rated at a maximum data transfer rate of 56Kb/s.

(2) ADSL Connection

Another type of dial-up connection is ADSL which is the most common way to access the Internet today. Sometimes ADSL is called as broadband accessing, because it provides wider bandwidth than that the regular modem provides.

ADSL was first developed in the 1980s the telecommunications industry's answer to the cable industry' request to support video on demand. In the middle 1990s, however, it was quickly recognized as a viable technology to enable access to high-speed services such as the

Internet[12]. ADSL delivers asymmetric transmission rates typically up to 9Mb/s downstream (from the CO to the premises) and 16Kb/s to 640Kb/s upstream (from the premises to the CO) as shown in Figure 3-5. Like all copper transmission systems, the higher the bit rate, the shorter the range. A limitation of ADSL transmission is distance: It can only be used within three miles of a telephone switching station, and the speed degrades as the distance gets closer and closer to the three-mile limit.

(3) Dedicated Connection

Unlike dial-up connections that only connect to your ISP's computer when you need to access the Internet, dedicated connections keep you continually connected to the Internet. With a dedicated connection, your PC is typically issues a static (non-changing) IP address to be used to transfer data back and forth via the Internet.

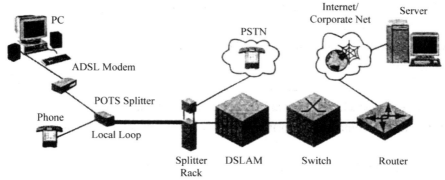

Figure 3-5 ADSL architecture

Types of dedicated Internet connections include connecting through a school or office LAN, as well as ADSL, cable, satellite, and fixed wireless connections.

(4) Wireless Connection

Wireless connections do not use a solid substance to connect sending and receiving devices. Rather, they use the air itself. Primary technologies used for wireless connections are infrared, broadcast radio, microwave, satellite and mobile wireless connections.

① Infrared

Infrared uses infrared light waves to communicate over short distances. It is sometimes referred to as line-of-sight communication because the light waves can only travel in a straight line. This requires that sending and receiving devices must be in clear view of one another without any obstructions blocking that view. One of the most common applications is to transfer data and information from a portable device such as a notebook computer or PDA to a desktop computer.

② Broadcast radio

Broadcast radio uses radio signals to communicate with wireless devices. For example, cellular telephones and many Web-enabled devices use broadcast radio to place telephone calls and/or to connect to the Internet [13]. Some end users connect their notebook or handheld computers to a cellular telephone to access the Web from remote locations. Most of these Web-enabled devices follow a standard known as Wi-Fi (wireless fidelity). This wireless standard is widely used to connect computers to each other and to the Internet.

③ Microwave

Microwave communication uses high-frequency radio waves. Like infrared, microwave communication provides line-of-sight communication because microwaves travel in a straight line. Because the waves can be transmitted only over relatively short distances, thus, microwave is a good medium for sending data between buildings in a city or on a large college campus.

Bluetooth is a short-range wireless communication standard that uses microwaves to transmit data over short distances of up to approximately 33 feet. Unlike traditional microwaves, Bluetooth does not require line-of-sight communication. Rather, it uses radio waves that can pass through nearby walls and other nonmetal barriers.

④ Satellite

Satellite can be used to send and receive large volumes of data. Uplink is a term relating to sending data to a satellite. Downlink refers to receiving data from a satellite. The major drawback to satellite communication is that bad weather can sometimes interrupt the flow of data.

One of the most interesting applications of satellite communications is for global positioning. A network of 24 satellites owned and managed by the Defense Department continuously sends location information to earth. Global Positioning System (GPS) devices use that information to uniquely determine the geographical location of the device.

⑤ Mobile Wireless Connection

Unlike satellite and fixed wireless connections, which use a cable to connect the modem to some type of fixed transceiver, mobile wireless connections allow the device to be moved from place to place. Consequently, most handheld PCs and other mobile devices (like Web-enable cell phones) use a mobile wireless connection and access the Internet through the same wireless network as cell phones and messaging devices.

NOTES

[1] 长句。句首 It 代表 Internet，since 引导的是原因状语从句，而从句中又包含两个由 which 引导的非限定性定语从句。

[2] to centralize…and to automate…为状语，后面都有介词短语 of…构成的定语，分别修饰两个 the task。此句说明 DNS 的功能是将网络中的主机名翻译成它的 IP 地址。

[3] HOSTS 文件，也是用于将主机名或者正式域名解析为相关的 IP 地址的，功能与 DNS 一样，只不过 HOSTS 比 DNS 出现得早，且早已被 DNS 取代了。SRI NIC 即斯坦

福研究院网络信息中心。

[4] 第一个 that 引导的是宾语从句；后面的 so that 引导的是目的状语从句。

[5] 最高层的域名是顶级域名，以行业为顶级域名的如.com（商业），.EDU（教育）等是指美国本土的，其他则以国家和地区为顶级域名，如 cn（中国）、HK（中国香港）等。

[6] 句中 it 代表 local DNS，until 引导的时间状语从句中，that can resolve…为定语从句，修饰 DNS server。provided…为状语，其中从属连接词 that 引导的是宾语从句，从句中 it 为先行代词，真正的主语为 a valid name。

[7] you might be familiar with 为定语从句，修饰 domains。

[8] that 引导的是宾语从句，句中 at the top 为主语；with the com through cn domains 为状语，underneath 副词，在底下之意。

[9] your modem dials up 为主动语态，实际上是对 modem 进行拨号。后一个 modem 是 ISP 机器上的。attached to…分词短语作定语。

[10] While 引导的是时间状语从句，为被动语态，实际上是完成连接之后，由 ISP 给 PC 指派一个临时 IP 地址。

[11] in your place 指所使用的 IP 地址，此句与上一句的意思紧密相关。

[12] to enable…为实际主语，it 为先行主语。

[13] cellular telephone 即移动电话。

KEYWORDS

access provider	访问提供商，接入提供商
ISP (Internet Service Provider)	因特网服务提供商
DNS (Domain Name System)	域名系统
domain name service	域名服务
host	主机
top level domain	顶级域
domain name server	域名服务器
client	客户
root domain	根域
sub-domain	子域
dial-up	拨号
modem	调制解调器
interface device	接口设备
IP(Internet Protocol)	网际协议，网间协议，互联网协议
transfer rate	传输速率
ADSL(Asymmetric Digital Subscriber Line)	非对称数字用户线
broadband	宽（频）带

VOD (Video On Demand)	视频点播
downstream	下行，顺流，下游
upstream	上行，逆流，上游
CO(Central Office)	中心（交换）局
dedicated connection	专线连接
LAN(Local Area Network)	局域网
satellite	卫星
wireless	无线
infrared	红外线
microwave	微波
line-of-sight communication	视距通信
portable device	便携设备
PDA(Personal Digital Assistant)	个人数字助理
Cellular telephone, cell phone	蜂窝电话
Web	万维网
end user	最终用户
Wi-Fi (Wireless Fidelity)	无线保真度（无线局域网）联盟
uplink	上行链路
downlink	下行链路
global positioning	全球定位

EXERCISES

Multiple Choices

1. The Internet is _____.
   a. a network of networks
   b. a public network
   c. the largest and most well-known computer network in the world
   d. a private network

2. DNS is _____.
   a. an abbreviation for Domain Name System
   b. used to translate network name to address assignments
   c. used at SRI NIC only
   d. the successor of HOSTS file

3. DNS has _____.
   a. a flat distributed architecture   b. a root server
   c. a hierarchical distributed architecture d. many root servers

4. Top-level domains in the DNS _____.

a. have two groups
b. have three groups
c. are located on the second layer of a similar inverted tree
d. don't include domain name of China
5. Following device belongs to the one of dial-up connection _____.
   a. Wi-Fi   b. modem
   c. dedicated line device   d. ADSL
6. When we use a dial-up connection to access the Internet, we need _____.
   a. a TV cable   b. a regular telephone line
   c. a modem   d. a temporary IP address
7. ADSL works at a rate of _____.
   a. asymmetric transmission
   b. symmetric transmission
   c. asymmetric transmission with 9Mb/s downstream
   d. asymmetric transmission with 16K-64Kb/s upstream
8. With the dedicated connection to connect the Internet we should _____.
   a. use dial-up connection   b. keep a continual connection
   c. use a static IP address   d. keep a temporary connection
9. Following technologies are used for wireless communication: _____.
   a. mobile wireless connection b. microwave
   c. satellite   d. broadcast radio
10. Infrared connection is used for _____.
    a. long distance communication
    b. short distance communication
    c. line-of-sight communication
    d. transferring data between a notebook computer and a desktop computer in one room
11. Using broadcast radio connection, we should use _____.
    a. Wi-Fi standard   b. Web-enabled devices
    c. mobile telephones   d. handheld computers
12. Bluetooth _____.
    a. is one type of microwave communication
    b. requires line-of-sight communication
    c. can pass through nonmetal barriers
    d. can be used to transmit data over short distance of up to approximately 10 meters
13. Satellite communication has _____.
    a. an uplink and a downlink
    b. a limitation of its application as bad weather can interrupt the flow of data
    c. an interesting application which is GPS

        d. the capability for sending and receiving large volumes of data
14. Mobile wireless connection _____.
        a. should be used to connect fixed devices
        b. should be used to connect mobile devices
        c. can be used to access Internet
        d. allows the device to be moved from place to place

PART II

INTERNET APPLICATIONS

CHAPTER 4
TRADITIONAL INTERNET APPLICATIONS

4.1 OVERVIEW OF WORLD WIDE WEB (WWW)

4.1.1 ABOUT WWW

WWW or Web is a large network of Internet servers providing hypertext and other services to terminals running client applications such as a browser [1].

WWW enables users to search, access, and download information from a worldwide series of networked servers where information is dynamically interlinked. A Web client passes a user's request for information to a server, usually by way of a Web browser. The server and client communicate through a transfer protocol, usually the HyperText Transfer Protocol (HTTP) [2]. The server then accesses a Web page using a Uniform Resource Locator (URL). Search engines are available to simplify access by enabling users to enter search criteria on a topic and have several URLs returned for Web pages that pertain to the desired information [3].

1. Providers

The most common way to access the Internet is through an Internet service provider (ISP). The providers are already connected to the Internet and provide a path or connection for individuals to access the Internet. Your college or university most likely provides you with free access to the Internet either through its local area networks or through a dial-up or telephone connection [4]. There are also some companies that offer free Internet access.

The most widely used commercial Internet service providers are national providers (e.g. America Online (AOL)) and wireless providers.

2. Browsers

Browsers are programs that provide access to Web resources. This software connects you to remote computers, opens and transfers files, displays text and images, and provides in one tool an uncomplicated interface to the Internet and Web documents. Browsers allow you to explore, or to surf the Web by easily moving from one Web site to another[5]. Three well-known browsers are Mozilla Firefox, Netscape Communications, and Microsoft Internet Explorer (See Figure 4-1).

Figure 4-1   Internet Explorer

Browser is a GUI-based hypertext client application, used to access hypertext documents and other services located on innumerable remote servers throughout the WWW and Internet.

As you see from Figure 4-2, what goes on when you click a hyperlink is a pretty significant series of events, involving not only your Web browser software, but also a Web server somewhere, and the transactions involved rely heavily on the HTML language.[6]

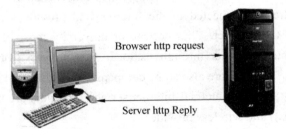

Figure 4-2   Web browser/server communication

3. Personal Web Sites

Do you have something to share with the world? Would you like a personal Web site, but don't want to deal with learning HTML? Creating your own home on the Internet is easy, and there are many services available to get you started[7].

A service site on the Web provides access to tools to create personal Web pages[8]. After registering with the site, you create your Web pages using the tools provided. Once completed, the service site acts as a host for your personal Web site and others are free to visit it from anywhere in the world.

NOTES

  [1] terminals 是指连入因特网的各种终端。

  [2] usually 后面是同位语。HTTP（HyperText Transfer Protocol）即超文本传输协议，是互联网上应用最为广泛的一种网络协议。所有的 WWW 文件都必须遵守这个标准。

  [3] 由 and 连接的是两个并列成分，主语为 Search engines；search criteria 是指搜索判据（关键字）。

  [4] most likely 意为"很可能"；either through…or through…方式状语。dial-up or telephone connection 是指用 modem 或 ADSL 方式连接，有关这方面的内容请参阅 3.2 节。

  [5] explore 是探索，surf 是冲浪，即在因特网上查看资料的行为。

  [6] 本段就是一句话。As 引导的是方式状语从句。主句中 what 引导的是主语从句，involving…现在分词短语作状语，其结构为 not only…but also…。

  [7] home 是指主页 homepage。

  [8] access to 此处为"使用"。

KEYWORDS

WWW (World Wide Web)	万维网，环球信息网
HyperText	超文本
client	客户
browser	浏览器
search	搜索
access	访问
download	下载
server	服务器
protocol	协议
HTTP (HyperText Transfer Protocol)	超文本传输协议
web page	网页
URL (Uniform Resource Locator)	统一资源定位符（地址）
search engine	搜索引擎
search criteria	搜索条件
ISP (Internet Service Provider)	因特网服务提供商
path	通路，路径
AOL(America On Line)	美国在线
explore	探索
surf	冲浪
personal Web site	个人网站
register	注册

| host | 宿主机，主机 |

EXERCISES

Multiple Choices

1. Web_____.
   a. provides hypertext services
   b. allows the user's terminals running server programs to be used
   c. allows the user's terminals running client programs to be used
   d. allows the user's terminals running browser to be used

2. A Web client_____.
   a. passes a user's request for information to another client
   b. passes a user's request for information to a server
   c. uses HTTP to transfer its request
   d. can use search engines

3. A Web server_____.
   a. accesses a Web page using URL
   b. can communicate with client via HTTP
   c. has several search engines to be used
   d. can send a search result to a client

4. ISP_____.
   a. is the abbreviation of Internet Service Provider
   b. can provide a connection for individuals to access the Internet
   c. has always to connect the Internet
   d. usually offers you with free access to the Internet

5. Browsers_____.
   a. are programs that provide access to Web resources
   b. provide a simple interface to the Web documents
   c. allow you to surf the Web
   d. connect you to remote computers

6. Creating our own Website_____.
   a. is difficult
   b. is easier
   c. can use some creating tools
   d. can ask some service sites to give helps

7. A service site on the Web_____.
   a. offers some tools to create personal Web pages
   b. should ask you to register to this site before using the tools of creating personal Web pages
   c. can act as a host for your personal Web site
   d. allows other sites to visit the personal Web site freely from anywhere in the world

8. Browser is _____.
   a. a GUI-based hypertext client application
   b. a GUI-based hypertext server application
   c. used to access hypertext documents
   d. used to access other services located on innumerable remote servers throughout the Web
9. To create a personal Web site, we _____.
   a. must learn HTML   b. don't need to learn HTML
   c. need some tools for creating the Web site   d. should choose a server site to be a host
10. A Web browser wants to communicate with a server via Internet _____.
    a. the browser should send a http request   b. the browser should send a html request
    c. the server should return a http reply   d. the server should return a html reply

4.1.2 SEARCH ENGINES

1. Overview

Imagine walking into a library where books are piled up and strewn about without any order [1]. Finding what you are looking for would be next to impossible [2]. As the World Wide Web rapidly grew, it is necessary to categorize message of Web page and to keep track of "what's what" and "what's where". In the initial stages of the Web, it was hard to locate useful information.

In the early days of the World Wide Web, two graduate students at Stanford University, Jerry Yang and David Filo, came up with a way to organize hyperlinks by category and they found it useful. In late 1993 this way was known as "Jerry Yang's Guide to the WWW". The name was soon changed to Yahoo! and the first search tool was born.

Today there is quite a collection of search tools available that allows us to find required information on the Web quickly and easily. The collection of search tools is constantly evolving, with new ones coming on the scene and others disappearing. Rather than report on how each of the popular search tools works, we will explain a few of them and suggest some Web presentations that provide reviews of all of the current search tools available [3].

2. How search engines work

A search engine operates, in the following order
(1) Web crawling
(2) Indexing [4]
(3) Searching

Web search engines work by storing information about many web pages, which they retrieve from the html itself. These pages are retrieved by a Web crawler (sometimes also known as a spider) — an automated Web browser which follows every link on the site[5]. The contents of each page are then analyzed to determine how it should be indexed (for example, words are extracted from the titles, headings, or special fields called metatags)[6]. Data about web pages are stored in an index database for use in later queries. A query can be a single word. The purpose of an index is to allow information to be found as quickly as possible. Some search engines, such as Google, store all or part of the source page (referred to as a cache) as well as information about the web pages, whereas others, such as AltaVista, store every word of every page they find. This cached page always holds the actual search text since it is the one that was actually indexed, so it can be very useful when the content of the current page has been updated and the search terms are no longer in it[7].

When a user enters a query into a search engine (typically by using key words), the engine examines its index and provides a listing of best-matching web pages according to its criteria, usually with a short summary containing the document's title and sometimes parts of the text[8].

NOTES

[1] where 引导的定语从句修饰 library；pile up 堆积。strewn 是 strew 的过去分词，意为"撒布"，about 此处为"到处"之意。

[2] what 引导的是宾语从句；next to 用于否定句之前，意为"几乎"。

[3] Rather than 引导方式状语从句，表示"而不是……"；how 引导的是宾语从句；that 引导的是定语从句，修饰 presentations，that 在从句中作主语。

[4] Indexing，搜索引擎单独控制它们的索引和排名算法。

[5] crawler，网上浏览器；破折号后面为同位语；which 引导的是定语从句。

[6] title 和 heading 均是标题；meta tag，元标签，是用于插入到网页首部的 HTML 代码。

[7] since 引导的是原因状语从句，it 代表主句中的 cached page，one 是指 page；that 引导的定语从句，修饰 one；so 引导的是结果状语从句，最后的 it 仍然是 cached page。

[8] 长句。When 引导的是时间状语；usually with…为状语；containing…分词短语作定语，修饰 summary。

KEYWORDS

search tool	搜索工具
spider	网络蜘蛛，万维网查询工具
index	索引
index database	索引数据库
cache	高速缓存
query	查询
key word	关键字

crawler 爬行者，浏览器

EXERCISES
True/False

1. _____ As the WWW rapidly grew, it is hard to find useful information.
2. _____ Google is the first search tool in the world.
3. _____ Collection of search tools is constantly evolving, with new ones coming on the scene and others disappearing.
4. _____ In order to operate a search engine, we should follow five operations.
5. _____ Search engines retrieve with http itself.
6. _____ Web crawler is an automated Web browser.
7. _____ Index database stores data concerning Web pages.
8. _____ Google stores every word of every page it finds.
9. _____ Cached page always holds the actual search text.
10. _____ A personal Web page is very useful when the content of its current page has been updated.
11. _____ When we enter a query into a search engine, the engine examples its index and provides a listing of best-matching Web pages.
12. _____ Purpose of an index is to allow information to be found as quickly as possible.

4.2 E-MAIL

E-mail or electronic mail is the transmission of electronic messages over the Internet. At one time, E-mail consisted only of basic text messages. Now E-mail routinely includes graphics, photos, and many different types of file attachments. People all over the world send E-mail to each other. You can E-mail your family, your co-workers. All you need to send and receive E-mail is an E-mail account, access to the Internet, and an E-mail program[1]. Two of the most widely used E-mail programs are Microsoft's Outlook Express and Mozilla Thunderbird.

A typical E-mail message has three basic elements: header, message, and signature (See Figure 4-3). The header appears first and typically includes the following information:

Business Email Sample
To: "Anna Jones" <annajones@buzzle.com<
Cc: All Staff
From: "James Brown"
Subject: Welcome to our Hive!

Figure 4-3 Basic elements of an E-mail message

(1) Addresses

Addresses of the persons sending, receiving, and, optionally, anyone else who is to receive copies. E-mail addresses have two basic parts (See Figure 4-4). The first part is the user's name and the second part is the domain name, which includes the top-level domain. In our example E-mail, dcoats is Dan's user name. The server providing E-mail service for Dan is usc.edu. The top-level domain indicates that the provider is an educational institution[2].

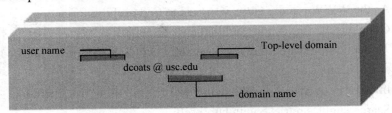

Figure 4-4　Two parts of an E-mail address

(2) Subject

A one-line description, used to present the topic of the message. Subject lines typically are displayed when a person checks his or her mailbox.

(3) Attachments

Many E-mail programs allow you to attach files such as documents and worksheets. If a message has an attachment, the file name appears on the attachment line.

The letter or message comes next. It is typically short and to the point. Finally, the signature line provides additional information about the sender. Typically, this information includes the sender's name, address, and telephone number.

E-mail can be a valuable asset in your personal and professional life. However, like many other valuable technologies, there are drawbacks too. Americans receive billions of unwanted and unsolicited e-mails every year. This unwelcome mail is called spam. While spam is indeed a distraction and nuisance, it also can be dangerous. For example, computer viruses or destructive programs are often attached to unsolicited E-mail.

In an attempt to control spam, anti-spam laws have been added to our legal system. For example, the recently enacted CAN-SPAM Act requires that every marketing-related e-mail provide an opt-out option[3]. When the option is selected, the recipient's e-mail address is to be removed from future mailing lists. This approach, however, has had minimal impact since over 50 percent of all spam originates from servers outside the United States[4]. A more effective approach has been the development and use of spam blockers (See Figure 4-5). These programs use a variety of different approaches to identify and eliminate spam.

Spam Blocker	Site
InBoxer	www.inboxer.com
OnlyMyEmailPersonal	www.onlymymail.com
Qurb	www.qurb.com
Vanquish vqME	www.vanquish.com

Figure 4-5　Spam blockers

NOTES

[1] you need…定语从句，修饰 All；is 后面有 3 个并列的表语。

[2] that 引导的是宾语从句。

[3] opt-out，退出。

[4] since 此处作副词用，为"其后""从那时起"之意。

KEYWORDS

E-mail	电子邮件
electronic message	电子消息（报文，电文）
attachment	附件
account	账户（号）
header	头部
signature	签名
address	地址
user's name	用户名
domain name	域名
subject	主题
spam	垃圾邮件
mailing list	邮件地址表，邮件清单
spam blocker	垃圾邮件拦截器

EXERCISES

Fill in the blanks with using terms, words or phrases found behind this exercises.

1. E-mail is the transmission of electronic messages over the _____.
2. Now E-mail routinely includes _____.
3. An E-mail account, access to the Internet, and an E-mail program are _____.
4. A typical E-mail message has three basic elements: _____.
5. The header of an E-mail message includes the following information: _____.
6. E-mail addresses include two basic parts: _____.
7. _____ is used to present the topic of a message.
8. Documents and worksheets can be used to be _____.
9. E-mail can be a _____ in your personal and professional life.
10. The unwanted and unsolicited E-mail are called _____.
11. _____ are often attached to unwelcome E-mail.
12. In our legal system _____ have been added.
13. If we want to remove our E-mail address from future mailing lists, we should select an _____.

14. _____ use a variety of different ways to identify and eliminate spam.

    a. spam

    b. addresses, subject and attachments

    c. Internet

    d. Subject

    e. Spam blockers

    f. all you need to send and receive E-mail

    g. valuable asset

    h. anti-spam laws

    i. graphics, photos, and many different types of file attachment

    j. user's name and domain name

    k. opt-out option

    l. Computer viruses or destructive programs

    m. header, message, and signature

    n. attachments

4.3 ELECTRONIC COMMERCE AND THE INTERNET OF THINGS

4.3.1 ELECTRONIC COMMERCE

1. What is Electronic Commerce

Electronic commerce is a system that includes not only those transactions that center on buying and selling goods and services to directly generate revenue, but also those transactions that support revenue generation, such as generating demand for those goods and services, offering sales support and customer service (See Figure 4-6), or facilitating communications between business partners [1].

Electronic commerce builds on the advantages and structures of traditional commerce by adding the flexibilities offered by electronic networks.

Electronic commerce enables new forms of business, as well as new ways of doing business. Amazon.com, for example, is a bookseller based in Seattle, Washington. The company has not physical stores, sells all their books via the Internet, and coordinates deliveries directly with the publishers so they do not have to maintain any inventory [2]. Companies such as Kantara and software.net take this a step further.

Because all of their products (commercial software packages) are electronic, and can be stored on the same computers that they use for processing orders and serving the Web, their inventory is totally digital [3]. As another example, AMP Inc. is offering its clients the opportunity to purchase electronic connectors and related components directly from its Web-based catalog, bypassing the need for EDI-based purchase orders and confirmations [4].

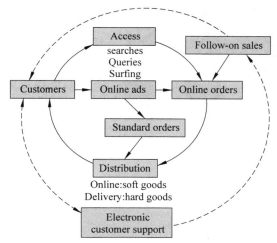

Figure 4-6　Cycle of electronic commerce

2. E-Commerce Business Models

A company's policies, operations, and technology define its business model. In essence, a company's business model describes how the company generates revenue. There are a number of standard e-commerce business models, including business to consumer (B2C), business to business (B2B), consumer to consumer (C2C), and business to government (B2G). Some of the most commonly used models are discussed next.

(1) Business to Consumer (B2C)

With the business-to-consumer (B2C) model, businesses sell goods or services to individual consumers. The B2C model was one of the first major types of e-commerce business models to be defined and implemented using the Web. Some examples of B2C businesses include Amazon.com, L.L. Bean, Walmart.com, and Polo.com (See Figure 4-7). These businesses can be Internet-only stores, or click-and-mortar stores with both online and traditional storefronts [5].

(2) Business to Business (B2B)

Business-to-business (B2B) applications include any type of e-commerce transaction taking place between two businesses. B2B revenue is increasing and expected to continue to grow tremendously in the next few years.

(3) Consumer to Consumer (C2C)

The consumer-to-consumer (C2C)—sometimes referred to as the person-to-person or P2P—business model almost solely consists of consumer auctions, where consumers sell products to other consumers [6]. With millions of products for sale every day, eBay is one of the largest C2C e-commerce business today.

Figure 4-7　Business to consumer

(4) Business to Government (B2G)

With U.S. government spending exceeding $500 billion per year with no signs of slowing down, business-to-government (B2G) organizations are becoming more prominent[7]. These organizations sell products and services to local, state, and federal government buyers. In general, the government sector has been slower to embrace online buying than the private sector.

Related B2G activities include some states allowing citizens to make payments online, such as paying taxes, renewing driver's licenses, and so forth [8]. This is sometimes referred to as C2G or customer-to-government e-commerce.

NOTES

[1] 这一段就是一句话，其中第一个 that 引出了 not only…，but also…结构的定语从句，在此从句中又有两个 that 引导的定语从句，分别修饰各自的 transactions；such as 后面为同位语，此同位语由 3 个分词短语构成。

[2] 本句有三个谓语：has、sells 和 coordinates；so 引导的是结果状语从句。

[3] 长句。前面大部分是 Because 引导的原因状语从句，仅最后的 their…为主句。

[4] bypassing，此处可译为"无须……"。

[5] click-and-mortar，混合经营的。

[6] where 引导的是非限定性定语从句。

[7] with 引出的短语作状语。

[8] allowing citizens…现在分词短语作定语。

KEYWORDS

electronic commerce	电子商务
B2B(B to B, Business-to-Business)	商业对商业
business mode	企业模式，业务模式
B2C (Business-to-Consumer)	企业对消费者
C2C (Consumer-to-Consumer)	消费者对消费者
B2G (Business-to-Government)	企业对政府
P2P (Person-to-Person)	个人对个人
C2G (Customer-to-Government)	客户对政府
supply chain	供应链
click-and-mortar	混合经营的

EXERCISES

Multiple Choices

1. Electronic Commerce supports _____.
    a. buying goods
    b. selling goods
    c. services of directly generating revenue
    d. other transactions
2. Electronic Commerce is _____.
    a. an EDI-based system only
    b. built on the structures of traditional commerce
    c. used with electronic networks
    d. built on the advantages of traditional commerce
3. The company that runs electronic stores has _____.
    a. not physical store   b. one physical store at least
    c. a digital inventory   d. a Web site
4. Electronic Commerce _____.
    a. provides new forms of doing business
    b. does not have to maintain any inventory
    c. is a system that offers customer service
    d. provides new ways of doing business
5. Business model can be defined by a company's _____.
    a. policies   b. operations   c. technology   d. revenues
6. B2C means _____.

a. businesses sell goods to individual consumers
   b. business-to-consumer
   c. consumers sell service to businesses
   d. individuals buy goods from businesses
7. C2C means _____.
   a. consumer sells auctions to other consumers
   b. consumer sells auctions to individuals
   c. consumer-to-consumer
   d. P2P
8. With B2G, citizens can make _____ online.
   a. paying taxes   b. watching TV
   c. renewing driver's licenses d. payments

4.3.2 INTERNET OF THINGS

1. Overview of The Internet of Things

The Internet of Things (IoT) is the network of physical objects or "things" embedded with electronics, software, sensors and connectivity to enable it to achieve greater value and service by exchanging data with the manufacturer, operator and/or other connected devices[1]. Each thing is uniquely identifiable through its embedded computing system but is able to interoperate within the existing Internet infrastructure.

Typically, IoT is expected to offer advanced connectivity of devices, systems, and services that goes beyond machine-to-machine communications (M2M) and covers a variety of protocols, domains, and applications. The interconnection of these embedded devices (including smart objects), is expected to usher in automation in nearly all fields, while also enabling advanced applications like a Smart Grid[2].

Things, in the IoT, can refer to a wide variety of devices such as heart monitoring implants, biochip transponders on farm animals, electric clams in coastal waters, automobiles with built-in sensors, or field operation devices that assist fire-fighters in search and rescue. These devices collect useful data with the help of various existing technologies and then autonomously flow the data between other devices. Current market examples include smart thermostat systems and washer/dryers that utilize wifi for remote monitoring.

Besides the plethora of new application areas for Internet connected automation to expand into, IoT is also expected to generate large amounts of data from diverse locations that is aggregated at a very high-velocity, thereby increasing the need to better index, store and process such data[3].

2. Architecture of The Internet of Things

The IoT system is likely to have an event-driven architecture. In Figure 4-8, IoT development is shown with a three-layer architecture. The top layer is formed by driven applications. The application space of the IoT is huge. The bottom layers represent various types of sensing devices: namely RFID tags, ZigBee or other types of sensors, and road-mapping GPS navigators[4]. Signals or information collected at these sensing devices are linked to the applications through the could computing platform at the middle layer.

The signal processing clouds are built over the mobile networks, the Internet backbone, and various information networks at the middle layer. In the IoT, the meaning of a sensing event does not follow a deterministic or syntactic model. In fact, the service-oriented architecture (SOA) model is adoptable here[5]. A large number of sensors and filters are used to collect the raw data. Various compute and storage clouds and grids are used to process the data and transform it into information and knowledge formats. The sensed information is used to put together a decision-making system for intelligence applications. The middle layer is also considered as a Semantic Web[6]. Some actors (service, components, avatars) are self-referenced[7].

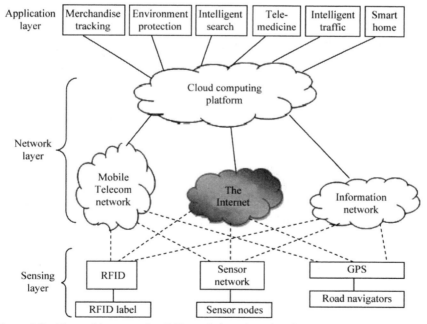

Figure 4-8 The architecture of an IoT consisting of sensing devices that are connected to various applications via mobile networks, the Internet, and processing clouds

3. Applications

According to Gartner, Inc. (a technology research and advisory corporation), there will be nearly 26 billion devices on the Internet of Things by 2020. ABI Research estimates that more than 30 billion devices will be wirelessly connected to the Internet of Things (Internet of Everything) by 2020[8]. As per a recent survey and study done by Pew Research Internet Project, a large majority of the technology experts and engaged Internet users who responded—83 percent—agreed with the notion that the Internet/Cloud of Things, embedded and wearable computing (and the corresponding dynamic systems) will have widespread and beneficial effects by 2025[9]. It is, as such, clear that the IoT will consist of a very large number of devices being connected to the Internet.

The ability to network embedded devices with limited CPU, memory and power resources means that IoT finds applications in nearly every field. Such systems could be in charge of collecting information in settings ranging from natural ecosystems to buildings and factories, thereby finding applications in fields of environmental sensing and urban planning. On the other hand, IoT systems could also be responsible for performing actions, not just sensing things. Intelligent shopping systems, for example, could monitor specific users' purchasing habits in a store by tracking their specific mobile phones. These users could then be provided with special offers on their favorite products, or even location of items that they need. Additional examples of sensing and actuating are reflected in applications that deal with heat, electricity and energy management, as well as cruise-assisting transportation systems.

However, the application of the IoT is not only restricted to these areas. Other specialized use cases of the IoT may also exist. An overview of some of the most prominent application areas is provided here. Based on the application domain, IoT products can be classified broadly into five different categories: smart wearable, smart home, smart city, smart environment, and smart enterprise. The IoT products and solutions in each of these markets have different characteristics.

NOTES

[1] 长句。embedded with…，过去分词短语作定语，修饰 objects or "things"。句中 to enable it…是目的状语，it 仍代表 objects or "things"。

[2] Smart Grid，智能电网，即电网智能化。

[3] 长句，主句是 IoT is also…，句中 that 引导的定语从句，修饰 data，thereby…结果状语。

[4] RFID（Radio Frequency Identification）即射频识别，又称无线射频识别，是通过无线电信号识别特定目标并读写相关数据的一种技术。在识别中，无须在识别系统与特定目标之间建立机械或光学接触。ZigBee 是一种近距离、低复杂度、低功耗、低成本的双向无线通信技术。GPS（Global Positioning System）即全球定位系统，起源于 1958 年美国军方的一个项目。到 1994 年，全球覆盖率达 98％的 GPS 卫星星座已布设完成。

[5] SOA（Service Oriented Architecture）model 即面向服务的体系结构模型，它将应用程序的不同功能单元（称为服务），通过定义良好的接口和协议联系起来，使得各种系

统的服务，可以使用一个统一的和通用的方式进行交互。

[6] Semantic Web，语义网，是 1998 年就提出来的一个概念，其核心是：通过给全球信息网上的文档（如 HTML）添加能够被计算机所理解的语义（元数据），从而使整个互联网成为一个通用的信息交换媒体。

[7] avatar，阿凡达，指在虚拟实境中互动地呈现一个人，即所谓的计算机化身。

[8] ABI Research，是一个技术市场智能公司，它收集所有公司的研究新闻，且每周都进行公布。Internet of Everything（IoE）称为万物互联。

[9] 长句。As per a recent survey…Project，为方式状语。主句为 a large majority of…agreed…，其中 that 引导的是宾语从句。

KEYWORDS

Internet of Things(IoT)	物联网
object	目标，对象，客体
embedded	嵌入（式）的
sensor	传感器
connectivity	连通性，连通度
identifiable	可辨识的，可视为同一的，可确认身份的
interoperate	（相）互操作
infrastructure	基础设施，基础结构
M2M (Machine-to-Machine)	机器对（到）机器
protocol	协议
domain	域，定义域，范畴
interconnection	互连，互接
biochip	生物芯片
transponder	发射应答机，转发器，询问机
thermostat	自动调温器
aggregate	集合，聚集，集合的，合计的
event-driven	事件驱动
RFID(Radio Frequency Identification)	射频识别，无线射频识别
GPS navigator	全球定位导航器
backbone	主干，骨干
syntactic	句法的
SOA (Service Oriented Architecture)	面向服务的体系结构
decision-making system	决策支持系统
avatar	具体化，具体表现，阿凡达
self-reference	自参考，自引用
wearable computing	可穿戴计算技术
ecosystem	生态系统

EXERCISES

True/False

1. _____ The things in IoT are embedded with electronics, software, and sensors.
2. _____ M2M communications go beyond IoT in offering advanced connectivity of devices systems, and services.
3. _____ Physical objects in IoT can refer to a wide variety of devices.
4. _____ Automobiles with built-in sensors are the devices that connect into Internet of Things.
5. _____ We do not need to better index, store and process volume data in IoT recently.
6. _____ IoT system has an event-driven architecture.
7. _____ IoT development has a four-layer architecture.
8. _____ The signal processing clouds are at the middle layer in the architecture of an IoT.
9. _____ In IoT the meaning of a sensing event should follow a deterministic or syntactic model.
10. _____ Various compute and storage clouds are used to process the raw data.
11. _____ By 2025 wearable computing is a widespread industry only.
12. _____ The ability to network embedded devices means that IoT finds applications in nearly every fields.
13. _____ The major task of IoT system is sensing only.
14. _____ Through tracking users' mobile phones the intelligent shopping systems could monitor their habits.
15. _____ IoT products can be classified into five categories.
16. _____ The Network layer in Figure 4-8 can be considered as a Semantic Web.

CHAPTER 5
NEW INTERNET APPLICATIONS

5.1 INSTANT MESSAGING (IM)

5.1.1 QQ

1. Overview of the QQ

Tencent QQ, generally referred to as QQ, is the most popular free instant messaging computer program in Mainland China. As of September 30, 2010, the active QQ user accounts for QQ IM amounted to 636.6 million, possibly making it the world's largest online community[1]. The number of simultaneous online QQ accounts exceeded 100 million. In February 2011, QQ.com ranked 10th overall in Alexa's internet rankings just behind Twitter ranked 9th[2]. The program is maintained by Tencent Holdings Limited (HKEX: 0700), owned in part by Naspers[3]. Since its entrance into Chinese households QQ quickly emerged as a modern cultural phenomenon, now being portrayed in popular culture[4]. Aside from the chat program, QQ has also developed many sub-features including games, virtual pets, ringtone downloads, music, shopping, blogs, microblogging, and group and voice chat etc.

The current version of QQ is QQ2010 beta2. Tencent periodically releases special versions of QQ to coincide with events such as the Olympics or Chinese New Year.

The official client runs on Microsoft Windows and a beta public version was launched for Mac OS X version 10.4.9 or newer[5]. The Web versions, WebQQ (full version) and WebQQ Mini (Lite version), which makes use of Ajax, are currently available[6].

As of January 2015, there are 829 million active QQ accounts, with a peak of 176.4 million simultaneous online QQ users.

2. QQ International

(1) Windows

In 2009 QQ began to expand its services internationally with its QQ International client for Windows distributed through a dedicated English-language portal.

QQ International offers non-Mandarin speakers the opportunity to use all the features of its Chinese counterpart to get in touch with other QQ users via chat and videocalls, it provides a non-Mandarin interface to access Qzone, Tencent's social network[7]. The client supports English, French, Spanish, German, Korean, Japanese and traditional Chinese. A wealth of third-party applications is bundled with QQ International and it is mainly aimed at making cross-cultural communications in and out of China more convenient.

One of the main features of QQ International is the optional and automatic machine translation of all chats.

(2) Android

An Android version of QQ International was released in September 2013. The client's interface is in English, French, Spanish, German, Korean, Japanese and traditional Chinese. In addition to text messaging, users can send each other images, video, and audio media messages. Moreover, users can share multimedia content with all contacts through the client's Qzone interface.

The live translation feature is available for all incoming messages and supports up to 18 languages.

(3) iOS / iPhone

QQ International for iPhone and iOS devices was released at the end of 2013, fully equivalent to its Android counterpart.

3. Web QQ

Tencent launched its web-based QQ formally on 15 September 2009, the latest version of which being 3.0. Rather than solely a web-based IM, WebQQ 3.0 functions more like its own operating system, with a desktop in which web applications can be added.

4. Open source and cross-platform clients

Using reverse engineering, open source communities have come to understand the QQ protocol better and have attempted to implement client core libraries compatible with more user-friendly clients[8]. Most of these clients are cross-platform, so they are usable on operating systems which the official client does not support. However, these implementations had only a subset of the functions of the official client and therefore were more limited in features.

NOTES

[1] making 现在分词短语作状语；it 代表 QQ，the world's…为宾语补语。

[2] Alexa 是一家专门发布网站世界排名的网站，同时提供网站的流量、网站访问量等。Twitter（非官方称为"推特"）是一个社交网在线服务网站，有关它的内容详见 5.1.3 节。

[3] Tencent Holdings Limited 是中国的一家国有公司，其子公司提供因特网和移动电话的增值服务和在线广告服务。HKEX（Hong Kong Stock Exchange）为香港股市交易。Naspers 是南非的一家跨国媒体公司，主营电子媒体（包括付费电视、因特网和即时消息用户平台，以及其他相关技术）和印刷媒体（包括出版、发行以及杂志、报纸和书籍的印刷，也提供私人教育服务）。

[4] Since 引导的是时间状语从句；portray，描绘。

[5] Mac OS X 是苹果电脑的操作系统。

[6] Ajax 是一组交叉式 Web 应用的开发方法，用于在客户端创建异步式 Web 应用。

[7] 长句，前半句中，第一个动词不定式 to use…为定语，修饰 opportunity。后一个动词不定式 to get in…作状语。Qzone 是腾讯计算机系统有限公司（港交所 00700）于 2005 年推出的一个网志系统，目前仍活跃在中国大陆。

[8] reverse engineering 即逆向工程（又称反向工程），是一种技术过程，即对一项目标产品进行逆向分析及研究，从而演绎并得出该产品的处理流程、组织结构、功能性能规格等设计要素，以制作出功能相近，但又不完全一样的产品。

KEYWORDS

blog	博客
social network	社交网
microblogging	微博
voice chat	语音聊天
portal	门户
video call	视频呼叫
multimedia	多媒体
counterpart	副本，复本，对应物
open source	开源
cross platform	跨平台
reverse engineering	逆（反）向工程

EXERCISES

Multiple Choices

1. QQ is the most popular free instant messaging in _____.
   a. Japan　　　　b. P.R.C　　　　c. Singapore　　　　d. Mainland China
2. QQ Instant Messaging _____.
   a. has user accounts that exceed 600 million
   b. is the world's largest online community
   c. has more than 100 million of user simultaneous on line
   d. has entered into Chinese households
3. QQ has the following sub-features:_____.
   a. chat　　　　b. virtual pets　　　　c. games　　　　d. blogs
4. Special versions of QQ are used for coinciding with _____.
   a. Christmas　　　　　　　　b. Thanksgiving Day
   c. Chinese New Year　　　　d. Olympics
5. Beta public version of QQ was launched for Mac OS X version _____.
   a. newer　　　　b. older　　　　c. Web QQ　　　　d. 10.4.9

6. QQ International for Windows offers _____ speakers to use all the features of its Chinese counterpart.
   a. English   b. Korean   c. non-Mandarin   d. Mandarin
7. QQ International for Android _____.
   a. has a server's interface in Classic Chinese
   b. can be used to send text messaging only
   c. can be used to share multimedia content with all contacts
   d. can support up to 18 languages in the live translation
8. The client core libraries are _____.
   a. compatible with user-friendly clients
   b. open source communities
   c. cross-platforms
   d. usable on operating systems which the official client does not support
9. Tencent Holdings Limited _____.
   a. is a Chinese national company
   b. has a branch company which offers Internet added services
   c. has a branch company which offers mobile added services
   d. has a branch company which offers online advertising services
10. Until now QQ International has _____.
    a. a Windows version   b. an Android version
    c. versions for iPhone and iOS   d. all the above

5.1.2 FACEBOOK

1. Features

Facebook is a social networking service and website launched in February 2004, operated and privately owned by Facebook Inc. As of June 2014, Facebook has more than 1.3 billion active users. Users may create a personal profile with photo, lists of personal interests, contact information, and other personal information, add other users as friends, and exchange messages, including automatic notifications when they update their profile[1]. Additionally, users may join common-interest user groups, organized by workplace, school or college, or other characteristics.

Facebook has a number of features with which users may interact:

① Wall, a space on every user's profile page that allows friends to post messages for the user to see [2].

② Pokes, which allows users to send a virtual "poke" to each other (a notification then tells a user that they have been poked)[3].

③ Photos, where users can upload albums and photos.

④ Status, which allows users to inform their friends of their whereabouts and actions.

Depending on privacy settings, anyone who can see a user's profile can also view that user's Wall.

On February 23, 2010, Facebook was granted a patent on certain aspects of its News Feed. The patent covers News Feed in which links are provided so that one user can participate in the same activity of another user. The patent may encourage Facebook to pursue action against websites that violate its patent, which may potentially include websites such as Twitter [4].

One of the most popular applications on Facebook is the Photos application, where users can upload albums and photos. Facebook allows users to upload an unlimited number of photos, compared with other image hosting services such as Photobucket and Flickr, which apply limits to the number of photos that a user is allowed to upload [5]. During the first years, Facebook users were limited to 60 photos per album. As of May 2009, this limit has been increased to 200 photos per album.

2. Technical aspects

Facebook is built in PHP which is compiled with HipHop for PHP, a "source code transformer" built by Facebook engineers that turns PHP into C++ [6]. The deployment of HipHop reportedly reduced average CPU consumption on Facebook servers by 50%.

Facebook used a combination platform based on HBase to store data across distributed machines [7]. Using a tailing architecture, new events are stored in log files, and the logs are tailed. The system rolls these events up and writes them into storage. The User Interface then pulls the data out and displays it to users. Facebook handles requests as AJAX behavior [8]. These requests are written to a log file using Scribe (developed by Facebook) [9].

NOTES

[1] 长句，主句为 Users may create…, add…, and exchange…，有 3 个并列的谓语；including…分词短语作状语。

[2] Wall 在 Facebook 中是指一块可供大家在上面留言的墙。

[3] poke，原意是"存入，插入"。此处是用手指戳戳的意思，是用来"烦"别人的。

[4] 句中有 that 和 which，分别引导了定语从句和非限定性定语从句。Twitter 的解释详见 5.1.3 节。

[5] 长句。compared with…过去分词短语作状语；which 引导的是非限定性定语从句；that 引导的是定语从句。Photobucket 是美国一个影像寄存、视频寄存、幻灯片制作与照片分享的网站。Flickr，雅虎旗下图片分享网站，提供免费及付费数码照片存储、分享方案之线上服务，也是网络社群服务的平台。

[6] PHP（Hypertext Preprocessor）即超文本预处理器，是一种通用开源脚本语言，详见 6.3 节。HipHop for PHP 是 Facebook 的一个项目，它由一个 PHP 到 C++的转换程序、

一个重新实现的 PHP 运行库和许多常用 PHP 扩展的重写版本构成，目的旨在加速和优化 PHP。

[7] HBase（Hadoop Database）是一个高可靠性、高性能、面向列、可伸缩的分布式存储系统，利用 HBase 技术可在廉价 PC Server 上搭建起大规模结构化存储集群。

[8] AJAX（Asynchronous JavaScript And XML）即异步 JavaScript 和 XML，是指一种创建交互式网页应用的网页开发技术。

[9] Scribe 是 Facebook 开源的日志收集系统，在 Facebook 内部已经得到大量的应用。它能够从各种日志源上收集日志，存储到一个中央存储系统（可以是 NFS，分布式文件系统等）上，以便于进行集中统计分析处理。它为日志的"分布式收集，统一处理"提供了一个可扩展的、高容错的方案。

KEYWORDS

profile	配置文件，概貌，轮廓
patent	专利
hosting	托管
upload	加载，上载，（向上）装入
compile	编译
consumption	消耗，耗散，消费
monolithic	单块（片）的
distributed machines	分布式（计算）机
log file	日志文件
aggregate	集合，聚集，集合的，合计的

EXERCISES

True/False

1. _____ Facebook is a social networking service.
2. _____ Users of Facebook can create a personal profile.
3. _____ Wall in Facebook is like a firewall.
4. _____ Poke in Facebook is a notification.
5. _____ With Photos of Facebook, we can download albums and photos.
6. _____ With Status of Facebook, users can locate their friends.
7. _____ Certain aspects of Facebook's NewsFeed are patent.
8. _____ Flickr allows users to upload unlimited number of photos.
9. _____ Facebook allows users to upload limited number of photos.
10. _____ Before 2005 Facebook users were limited to 60 photos per album.
11. _____ HipHop for PHP is a source code transformer.
12. _____ A report shows that the use of HipHop reduced average CPU consumption on Facebook servers by 40%.

13. _____ Using a tailing architecture, new events are stored in log files.
14. _____ HBase is a high reliable, high performance and resize-able distributed storage system.

5.1.3 TWITTER

Twitter is an online social networking service that enables users to send and read short 140-character messages called "tweets" [1].

Registered users can read and post tweets, but unregistered users can only read them. Users access Twitter through the website interface, SMS, or mobile device app. Twitter Inc. is based in San Francisco and has more than 25 offices around the world.

Twitter was created in March 2006. The service rapidly gained worldwide popularity, with more than 100 million users who in 2012 posted 340 million tweets per day. The service also handled 1.6 billion search queries per day. In 2013 Twitter was one of the ten most-visited websites, and has been described as "the SMS of the Internet." As of December 2014, Twitter has more than 500 million users, out of which more than 284 million are active users [2].

1. Features

Tweets are publicly visible by default, but senders can restrict message delivery to just their followers. Users can tweet via the Twitter website, compatible external applications (such as for smartphones), or by Short Message Service (SMS) available in certain countries[3]. Retweeting is when a tweet is forwarded via Twitter by users. Both tweets and retweets can be tracked to see which ones are most popular. While the service is free, accessing it through SMS may incur phone service provider fees.

Twitter allows users to update their profile via their mobile phone either by text messaging or by apps released for certain smartphones and tablets.

As a social network, Twitter revolves around the principle of followers. When you choose to follow another Twitter user, that user's tweets appear in reverse chronological order on your main Twitter page. If you follow 20 people, you'll see a mix of tweets scrolling down the page: breakfast-cereal updates, interesting new links, music recommendations, even musings on the future of education.

2. Implementation

Twitter places great reliance on open-source software. The Twitter Web interface uses the Ruby on Rails framework, deployed on a performance enhanced Ruby Enterprise Edition implementation of Ruby[4].

As of April 6, 2011, Twitter engineers confirmed they had switched away from their Ruby on Rails search stack, to a Java server they call Blender[5].

The service's application programming interface (API) allows other web services and applications to integrate with Twitter[6].

NOTES

[1] tweet，推文，tweet 原意是小鸟啾啾地叫，此处为短小消息。

[2] out of which，原意是出自……，which 是指 5 亿用户。

[3] tweet，此处为动词。 SMS(Short Message Service)，短信服务。

[4] Ruby on Rails 是一个可以使开发、部署、维护 Web 应用程序变得简单的框架。Ruby Enterprise Edition (REE)是标准 Ruby（一种为简单快捷的面向对象编程而开发的脚本语言）解析器的改进版本,号称能够让 Rails（是一个更符合实际需要而且更高效的 Web 开发框架）应用节约 1/3 的内存使用量，并且能更好地提高性能。

[5] confirmed 后面省略了 that，这是一个宾语从句。Blender 是一款开源的跨平台全能三维动画制作软件，提供从建模、动画、材质、渲染到音频处理、视频剪辑等一系列动画短片制作解决方案。

[6] 此处 service 是指 Blender。

KEYWORDS

register	注册，寄存器
post	邮政，邮寄，记入，位置
SMS(Short Message Service)	短消息服务
tweet	推文
retweet	转推文
profile	配置文件，简表，概貌，轮廓
stack	栈，堆栈
Java	一种程序设计语言

EXERCISES

Fill in the blanks with appropriate words or phrases found behind this exercises.

1. Twitter allows users to send and receive short _____ called "tweets".

2. Unregistered Twitter users can only read _____.

3. By way of _____ users can access Twitter.

4. Senders in Twitter can _____ message delivery to just their followers.

5. In order to see which persons are most popular, we can _____ tweets and retweets.

6. Twitter places great reliance on _____.

7. Twitter revolves around the principle of _____.

8. _____ allows other web services and applications to integrate with Twitter.

9. Ruby is an object-oriented _____ language.

10. Twitter allows users to update their _____ via their mobile phone.

a. scripting

b. API

c. 140-character messages

d. open-source software

e. website, interface, SMS, or mobile device app

f. profile

g. tweet

h. track

i. restrict

j. followers

5.1.4 WECHAT

WeChat is a mobile text and voice messaging communication service developed by Tencent in China, first released in January 2011. It is the largest standalone messaging app by monthly active users.

The app is available on Android, iPhone, BlackBerry, Windows Phone and Symbian phones, and there are also Web-based and OS X clients but these require the user to have the app installed on a supported mobile phone for authentication[1]. As of August 2014, WeChat has 438 million active users; with 70 million outside of China.

Users can register WeChat with Facebook account or sign up with phone number. WeChat currently supports phone numbers of more than 100 countries to register. Registration cannot be done directly through Tencent QQ[2]. But users can connect their WeChat account with Tencent QQ account after registering through phone number.

WeChat provides text messaging, hold-to-talk voice messaging, broadcast (one-to-many) messaging, sharing of photographs and videos, and location sharing. It can exchange contacts with people nearby via Bluetooth, as well as providing various features for contacting people at random if desired and integration with social networking services such as those run by Facebook and Tencent QQ[3]. Photographs may also be embellished with filters and captions, and a machine translation service is available.

WeChat supports users to register as a public account, which enables them to push feeds to subscribers, interact with subscribers and provide them with service[4]. By the end of 2014, number of Wechat public accounts had reached 8 million.

In China, Wechat public accounts have become a common service or promotion platform for government, news media and companies. Specific public account subscribers use the platform for services like hospital pre-registration, visa renewal or credit card service.

On 30 September 2014, WeChat 6.0 was launched with new features and functions include Sight capture and share.

According to GlobalWebIndex, WeChat is the fifth most used smartphone app worldwide and in August 2013, following Google Maps, Facebook, YouTube and Google+[5]. WeChat claimed it had 100 million registered international users which are achieved in only 3 months from 50 million registered international users. It also claimed 300 million registered Chinese users.

According to Xinhua, WeChat total users reached 600 million users worldwide in October 2013. In addition, approximately 30 percent of the total WeChat are users abroad.

NOTES

[1] 由 and 连接的并列句，前一句中的 The app 是指前一段中的 the largest standalone messaging app，即 WeChat，后一句中 there are…省略了 to be used by the app。Android 见 2.3 节。 BlackBerry，加拿大 RIM(Research In Motion Ltd)公司生产的黑莓手机。Windows Phone（WP）是微软发布的一款手机操作系统，它将微软旗下的 Xbox Live 游戏、Xbox Music 音乐与独特的视频体验集成至手机中。

Symbian 系统是塞班公司为手机而设计的操作系统。

[2] Tencent QQ，腾讯 QQ，详见 5.1.1 节。

[3] 长句。It 代表 WeChat。句子最后的 run by…过去分词短语，修饰 those。Facebook，详见 5.1.2 节。

[4] push feeds 在 SNS 和微博中都要用到 feed，在微信中 feed 也指每条微信里的新鲜事物。在 feed 架构中要用到推（push）拉（pull）模式。其中 push 模式是把一篇微信推送给所有关注的人。

[5] Global WebIndex 是一个市场研究公司，其业务主要是向广告行业提供客户数据。Google Maps（谷歌地图）是谷歌提供的电子地图服务，包括局部详细的卫星照片。此款服务可以提供含有行政区和交通以及商业信息的矢量地图、不同分辨率的卫星照片和可以用来显示地形和等高线地形的视图。You Tube 是一个共享视频的网站，该网站 2006 年 11 月被谷歌公司收购，现为谷歌公司的一个子公司。Google+即 Google Plus，一个社交网络和为谷歌服务的社区。

KEYWORDS

WeChat	微信
message	消息，信息，电文
register	注册，寄存器
sign up	签约，参加，入队
text messaging	文本消息
broadcast	广播
photograph	照片，相片
video	视频
Bluetooth	蓝牙

random	随机的
integration	集成，整合
social networking service	社交网络服务
embellish	美化，装饰，修饰
filter	过滤器，滤波器，筛选程序
caption	标题，图片说明，字幕
subscriber	用户

EXERCISES

Fill in the blanks with appropriate words or phrases found behind this exercises.

1. WeChat is a _____.
2. The app of WeChat should install on a _____ for authentication.
3. We can register WeChat with _____ .
4. User can't register WeChat directly with _____.
5. Using WeChat users can share _____ .
6. With WeChat we can contact _____ at random if desired.
7. If we registered WeChat as a public account, we can interact with _____ .
8. In China, WeChat _____ have become a promotion platform for government.
9. _____ possesses new features such as Sight capture and share.
10. Approximately _____ of the total WeChat are users abroad.
11. WeChat was developed by _____ .
12. _____ was produced by RIM company in Canada.

   a. public accounts

   b. photographs and videos

   c. mobile text and voice messaging communication service

   d. BlackBerry

   e. WeChat6.0

   f. subscribers

   g. supported mobile phone

   h. 30%

   i. people

   j. Tencent QQ

   k. Tencent in China

   l. Facebook

5.2 SOCIAL NETWORKING SERVICE (SNS)

5.2.1 WIKI

A wiki is a website that allows the creation and editing of any number of interlinked web pages via a web browser using a simplified markup language or a WYSIWYG text editor[1]. Wikis are typically powered by wiki software and are often used collaboratively by multiple users. Examples include community websites, corporate intranets, knowledge management systems, and note services[2]. The software can also be used for personal notetaking.

Wikis serve different purposes. Some permit control over different functions (levels of access). For example, editing rights may permit changing, adding or removing material. Others may permit access without enforcing access control.

The essence of the wiki concept is as follows:

① A wiki invites all users to edit any page or to create new pages within the wiki Web site, using only a plain-vanilla Web browser without any extra add-ons[3].

② Wiki promotes meaningful topic associations between different pages by making page link creation almost intuitively easy and showing whether an intended target page exists or not[4].

③ A wiki is not a carefully crafted site for casual visitors. Instead, it seeks to involve the visitor in an ongoing process of creation and collaboration that constantly changes the Web site landscape.

A single page in a wiki website is referred to as a "wiki page", while the entire collection of pages, which are usually well interconnected by hyperlinks, is "the wiki"[5]. A wiki is essentially a database for creating, browsing, and searching through information.

A defining characteristic of wiki technology is the ease with which pages can be created and updated. Generally, there is no review before modifications are accepted. Many wikis are open to alteration by the general public without requiring them to register user accounts. Sometimes logging in for a session is recommended, to create a "wiki-signature" cookie for signing edits automatically. Many edits, however, can be made in real-time and appear almost instantly online. This can facilitate abuse of the system. Private wiki servers require user authentication to edit pages, and sometimes even to read them.

A wiki is an application also, typically a web application, which allows collaborative modification, extension, or deletion of its content and structure. In a typical wiki, text is written using a simplified markup language (known as "wiki markup") or a rich-text editor. While a wiki is a type of content management system, it differs from a blog or most other such systems in that the content is created without any defined owner or leader, and wikis have little implicit structure, allowing structure to emerge according to the needs of the users [6].

The encyclopedia project Wikipedia is the most popular wiki on the public web in terms of page views, but there are many sites running many different kinds of wiki software. Wikis can serve many different purposes both public and private, including knowledge management, notetaking, community websites and intranets. Some permit control over different functions (levels of access). For example, editing rights may permit changing, adding or removing material. Others may permit access without enforcing access control. Other rules may also be imposed to organize content.

NOTES

[1] 长句。that 引导定语从句，修饰 website；via…using…均为状语。WYSIWYG 的原文是"What You See Is What You Get"，即所见即所得。

[2] intranet 为因特网内联网，或称内部网、内联网、内网，是一个使用与因特网同样技术的计算机网络，它通常建立在一个企业或组织的内部，并为其成员提供信息的共享和交流等服务，例如万维网服务、文件传输、电子邮件等，是一种企业内部各个部门相互连接而形成的内部网。

[3] add-ons，附件，附加。

[4] by making…and showing…，方式状语。

[5] which 引导的是非限定性定语从句。

[6] 长句。While 引导的是让步状语从句，表示虽然……。in that 后面的结果状语从句是并列的，由 and 连接。

KEYWORDS

interlink	相互链接
community website	团体网站
intranet	内联网
add-on	附件
landscape	景色，风景
hyperlink	超链接
database	数据库
searching	搜索
user account	用户账户
log	登录
cookie	网络跟踪器，"小甜饼"
authentication	验证，鉴别
markup language	标记（置标）语言
text editor	文本编辑程序，文本编辑器
implicit structure	隐式结构

emerge		出现，露出，显现，发生
encyclopedia		百科全书
access		访问，接入，存取

EXERCISES

True/False

1. _____ With a markup language we can create and edit a lot of interlinked Web pages within the Wiki site.
2. _____ Wikis serve different purpose.
3. _____ Levels of access in wiki mean different functions.
4. _____ In wiki users who edit or create Web pages must use extra add-ons.
5. _____ Wiki can make decision to determine whether an intended target page exists or not.
6. _____ A wiki is a carefully crafted site for casual visitors.
7. _____ A single page in a wiki Web site is referred to as a "wiki page".
8. _____ Term "the wiki" means that the entire collection of pages are usually well interconnected by hyperlinks.
9. _____ When altering an opened wiki, we must register user accounts.
10. _____ Private wiki servers require user authentication to edit pages.
11. _____ When a wiki is used as a content management system, it is the same as a blog.
12. _____ Wiki allows collaborative modification, extension, or deletion of its content and structure.
13. _____ Wiki markup language is a simplified markup one.
14. _____ When we use wiki to create a file, we need to define the file's owner or leader.
15. _____ Wikis have been allowed its structure to be emerged.
16. _____ Wikipedia is an encyclopedia project.
17. _____ Wikis can serve many different purposes in public only.
18. _____ Some of wikis should permit to change the editing rights.

5.2.2　BLOG AND MICROBLOG

1. Blog

A blog (a blend of the term web and log) is a type of website or part of a website. Blogs are usually maintained by an individual with regular entries of commentary, descriptions of events, or other material such as graphics or video. Entries are commonly displayed in reverse-chronological order. Blog can also be used as a verb, meaning to maintain or add content to a blog.

Most blogs are interactive, allowing visitors to leave comments and even message each other via widgets on the blogs and it is this interactivity that distinguishes them from other static websites [1].

Many blogs provide commentary or news on a particular subject; others function as more personal online diaries[2]. A typical blog combines text, images, and links to other blogs, Web pages, and other media related to its topic. The ability of readers to leave comments in an interactive format is an important part of many blogs. Most blogs are primarily textual, although some focus on art (art blog), photographs (photoblog), videos (video blogging), music (MP3 blog), and audio (podcasting)[3]. Microblogging is another type of blogging, featuring very short posts.

As of 16 February 2011, there were over 156 million public blogs in existence. On 20 February 2014, there were around 172 million Tumblr and 75.8 million WordPress blogs in existence worldwide [4]. According to critics and other bloggers, Blogger is the most popular blogging service used today, however Blogger does not offer public statistics. Technorati has 1.3 million blogs as of February 22, 2014 [5].

There are many different types of blogs, differing not only in the type of content, but also in the way that content is delivered or written.

2. Microblog

Microblog is a broadcast medium in the form of blog. A microblog differs from a traditional blog in that its content is typically smaller in actual file size[6]. Microblogs "allow users to exchange small elements of content such as short sentences, individual images, or video links".

As with traditional blog, microbloggers post about topics ranging from the simple, such as "what I'm doing right now," to the thematic, such as "sports cars"[7]. Commercial microblogs also exist, to promote websites, services and/or products, and to promote collaboration within an organization.

Some microblog services offer features such as privacy settings, which allow users to control who can read their microblogs, or alternative ways of publishing entries besides the web-based interface[8]. These may include text messaging, instant messaging, E-mail, or digital audio.

Microblog services have revolutionized the way information is consumed. It has empowered citizens themselves to act as sensors or sources of data which could lead to important pieces of information. People now share what they observe in their surroundings, information about events, and what their opinions are about certain topics, for example, government policies in healthcare[9].

Moreover, these services store various metadata from these posts, such as location and time. Aggregated analysis of this data includes different dimensions like space, time, theme, sentiment, network structure etc., and gives researchers an opportunity to understand social perceptions of people in the context of certain events of interest[10]. Microblogging also promotes authorship. On the microblogging platform Tumblr, the reblogging feature links the post back to the original creator [11].

Microblogging has the potential to become a new, informal communication medium, especially for collaborative work within organizations. Over the last few years, communication patterns have shifted primarily from face-to-face to online in email, IM, text messaging, and other tools. However, some argue that email is now a slow and inefficient way to communicate. For instance, time-consuming "email chains" can develop, whereby two or more people are involved in lengthy communications for simple matters, such as arranging a meeting [12]. The one-to-many broadcasting offered by microblogs is thought to increase productivity by circumventing this.

Users and organizations can set up their own microblog service: free and open source software is available for this purpose. Hosted microblog platforms are also available for commercial and organizational use.

NOTES

[1] 并列长句。allowing…分词短语作状语；后一句中 that 引导的是定语从句，修饰 interactivity。

[2] function 此处为动词，意为"尽职责，起作用"；others 代表另外一些博客。

[3] art blog，艺博；photoblog，影博；video blogging，有时简称 vlogging 或 vidding 或 vidblogging，这种微博的媒体形式是视频或网络电视。

podcasting 是一种数字媒体广播技术（非流式网络广播），而 podcast 这种数字媒体文件按插曲出售，通常也可以通过万维网联合组织（Web Syndication）下载。

[4] Tumblr（汤博乐）成立于 2007 年，是目前全球最大的轻博客网站，也是轻博客网站的始祖。Tumblr 是一种介于传统博客和微博之间的全新媒体形态，既注重表达，又注重社交，而且注重个性化设置，成为当前最受年轻人欢迎的社交网站之一。雅虎公司董事会 2013 年 5 月 19 日决定，以 11 亿美元收购 Tumblr。WordPress 是一种使用 PHP 语言开发的博客平台，用户可以在支持 PHP 和 MySQL 数据库的服务器上架设属于自己的网站。也可以把 WordPress 当作一个内容管理系统（CMS）来使用。

[5] Technorati 是一个著名的博客搜索引擎，截至 2007 年 4 月，Technorati 已经索引了超过 7000 万个博客站点。可以说，Technorati 已经成为世界上最重要的博客搜索引擎之一。

[6] in that 引导的是状语从句。

[7] As 引导的是比较状语，主句中 ranging…为现在分词短语作定语；句中有两个 such as 构成的同位语。sports cars，跑车，赛车。

[8] such as privacy settings，…or alternative ways…为同位语；which 引导的是非限定性定语从句，who 引导的是宾语从句。

[9] 两个 what 引导的都是宾语从句。

[10] 由 and 连接的长句，主语是 Aggregated analysis of…。

[11] reblogging，是博客中的一种机制，它允许用户转发另一个用户博客的内容，并指出哪个用户是信息源。

[12] whereby，此处的用法相当于 by which。

KEYWORDS

entry	条目，词条，入口，进入
commentary	评论，注释
reverse-chronological order	反时间顺序
interactive	交互式的
widget	小装置（窗口），窗体
art blog	艺博
photoblog	影博
video blogging	视博
podcasting	Pod 广播技术
blogger	博客版主
statistics	统计，统计数字，统计表，统计学
microblog	微博
aggregate file	聚合文件
post	邮件，邮箱
microblogger	微博版主
open source software	开源软件
instant messaging	即时消息
digital audio	数字音频
authorship	著述业，著者，来源
circumventing	阻遏，设法规避，欺诈，绕行
hosted	托管的

EXERCISES

Multiple Choices

1. Blog is _____.
   a. a type of website   b. maintained usually by an individual
   c. a blend of the term web and log   d. a part of a website
2. A blog consists of _____.
   a. regular entries of commentary   b. descriptions of events
   c. E-mail   d. graphics or video
3. Most blogs _____.

a. are similar to other static websites   b. allow visitors to leave comments
   c. are interactive   d. use widgets on the blogs
4. A typical blog links to _____.
   a. other blogs   b. text
   c. web pages   d. other media related to its topic
5. The following is belong to special blog _____.
   a. art blog   b. photo blog   c. video blogging   d. MP3 blog
6. Difference among blogs is in _____.
   a. the type of content   b. the way that content is delivered
   c. the way that content is written   d. the way of linking
7. Microblog _____.
   a. is a broadcast medium in the form of blog
   b. allows user to exchange short sentences
   c. allows user to exchange longer sentences
   d. allows user to exchange individual images
8. Commercial microblogs can _____.
   a. replace traditional blog
   b. promote websites
   c. promote collaboration within an organization
   d. promote services
9. Entries published by microblog include _____.
   a. text   b. instant messaging
   c. E-mail   d. digital audio
10. Microblog services _____.
    a. have empowered citizens themselves to act as sensors of data
    b. have revolutionized the way information is consumed
    c. can be set up by organizations
    d. can be the hosted microblog platforms
11. Metadata gathered from Microblog services includes _____.
    a. space   b. time   c. theme   d. network structure
12. Communication patterns have shifted to _____.
    a. face-to-face   b. online E-mail
    c. Instant Messaging   d. text messaging

5.3 CLOUD COMPUTING

Cloud computing is the delivery of computing as a service rather than a product, whereby shared resources, software, and information are provided to computers and other devices as a utility (like the electricity grid) over a network (typically the Internet)[1].

1. Overview

Cloud computing is a marketing term for technologies that provide computation, software, data access, and storage services that do not require end-user knowledge of the physical location and configuration of the system that delivers the services[2]. A parallel to this concept can be drawn with the electricity grid, wherein end-users consume power without needing to understand the component devices or infrastructure required to provide the service[3].

Cloud computing describes a new supplement, consumption, and delivery model for IT services based on Internet protocols, and it typically involves provisioning of dynamically scalable and often virtualised resources[4]. It is a byproduct and consequence of the ease-of-access to remote computing sites provided by the Internet. This may take the form of web-based tools or applications that users can access and use through a web browser as if the programs were installed locally on their own computers[5].

Cloud computing providers deliver applications via the internet, which are accessed from web browsers and desktop and mobile apps, while the business software and data are stored on servers at a remote location. In some cases, legacy applications (line of business applications that until now have been prevalent in thin client Windows computing) are delivered via a screen-sharing technology, while the computing resources are consolidated at a remote data center location; in other cases, entire business applications have been coded using web-based technologies such as AJAX[6].

At the foundation of cloud computing is the broader concept of infrastructure convergence (or Converged Infrastructure) and shared services[7]. This type of data center environment allows enterprises to get their applications up and running faster, with easier manageability and less maintenance, and enables IT to more rapidly adjust IT resources (such as servers, storage, and networking) to meet fluctuating and unpredictable business demand[8].

Most cloud computing infrastructures consist of services delivered through shared data-centers and appearing as a single point of access for consumers' computing needs[9].

The tremendous impact of cloud computing on business has prompted the federal United States government to look to the cloud as a means to reorganize their IT infrastructure and decrease their spending budgets. With the advent of the top government official mandating cloud adoption, many agencies already have at least one or more cloud systems online.

A cloud computing logical diagram shows in Figure 5-1.

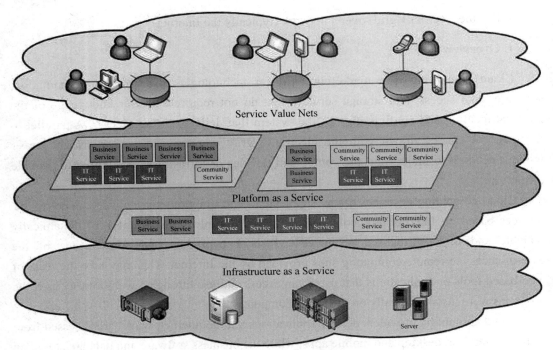

Figure 5-1　Cloud computing logical diagram

2. Public, Private, and Hybrid Clouds

A public cloud is built over the Internet and can be accessed by any user who has paid for the service. Public clouds are owned by service providers and are accessible through a subscription. The callout box in top of Figure 5-2 shows the architecture of a typical public cloud. Many public clouds are available, including Google App Engine (GAE), Amazon Web Services (AWS), Microsoft Azure, IBM Blue Cloud, and Salesforce.com's Force.com[10]. The providers of the aforementioned clouds are commercial providers that offer a publicly accessible remote interface for creating and managing VM instances within their proprietary infrastructure[11]. A public cloud delivers a selected set of business processes. The application and infrastructure services are offered on a flexible price-per-use basis.

A private cloud is built within the domain of an intranet owned by a single organization[12]. Therefore, it is client owned and managed, and its access is limited to the owning clients and their partners. Its deployment was not meant to sell capacity over the Internet through publicly accessible interfaces. Private clouds give local users a flexible and agile private infrastructure to run service workloads within their administrative domains. A private cloud is supposed to deliver more efficient and convenient cloud services. It may impact the cloud standardization, while retaining greater customization and organizational control. Intranet-based private clouds are linked to public clouds to get additional resources.

A hybrid cloud is built with both public and private clouds, as shown at the lower-left corner of Figure 5-2. Private clouds can also support a hybrid cloud model by supplementing local infrastructure with computing capacity from an external public cloud. For example, the Research Compute Cloud (RC2) is a private cloud, built by IBM, that interconnects the computing and IT resources at eight IBM Research Centers scattered throughout the United States, Europe, and Asia[13]. A hybrid cloud provides access to clients, the partner network, and third parties.

Figure 5-2 Public, private, and hybrid clouds illustrated by functional architecture and connectivity of representative clouds

In summary, public clouds promote standardization, preserve capital investment, and offer application flexibility. Private clouds attempt to achieve customization and offer higher efficiency, resiliency, security, and privacy. Hybrid clouds operate in the middle, with many compromises in terms of resource sharing.

3. Characteristics

Cloud computing exhibits the following key characteristics:

① Empowerment of end-users of computing resources by putting the provisioning of those resources in their own control, as opposed to the control of a centralized IT service.

② Agility improves with users' ability to re-provision technological infrastructure resources.

③ Application programming interface (API) accessibility to software that enables machines to interact with cloud software in the same way the user interface facilitates interaction between humans and computers[14]. Cloud computing systems typically use REST-based APIs[15].

④ Device and location independence enable users to access systems using a Web browser regardless of their location or what device they are using (e.g., PC, mobile phone). As infrastructure is off-site (typically provided by a third-party) and accessed via the Internet, users can connect from anywhere.

⑤ Multi-tenancy enables sharing of resources and costs across a large pool of users.

⑥ Reliability is improved if multiple redundant sites are used, which makes well-designed cloud computing suitable for business continuity and disaster recovery.

⑦ Performance is monitored, and consistent and loosely coupled architectures are constructed using web services as the system interface.

⑧ Security could improve due to centralization of data, increased security-focused resources, etc.

⑨ Maintenance of cloud computing applications is easier, because they do not need to be installed on each user's computer.

NOTES

[1] 长句，whereby 引导的是非限定性定语从句，whereby 的作用相当于 by which；as 引导的是方式状语。

[2] 长句，句中有三个 that 引导的定语从句。

[3] wherein 引导的是非限定性定语从句，wherein 的作用相当于 in which。

[4] 长句，由 and 连接的并列句；后一句中 it 代表 cloud computing；Internet protocols 实际上是指云计算是在因特网上实现的。

[5] This 代表 cloud computing；that 引导的定语从句，修饰 applications；as if…条件状语。

[6] 并列长句，中间由 while 连接，括号中 that 引导的定语从句，修饰 applications。thin client，瘦客户（lean client or slim client）是一种特别依赖于其他计算机（服务器）去完成通常的计算任务的计算机或程序。与其相对应的是胖客户（fat client），胖客户计算机则主要靠自己去完成计算任务。screen-sharing，屏幕共享，是苹果公司开发的，并

作为 VNC 客户机纳入 Mac OS X 10.5 版，只要激活屏幕共享程序即可远程观察和控制任何在局域网上的 Mac 机。AJAX 的解释详见 5.1.2 节 NOTES[8]。

　　[7] 此句为倒装句，At the foundation of cloud computing 是表语；主语是 the broader concept of…。infrastructure convergence 即基础会聚或称会聚的基础（converged infrastructure），是 IT 部门用来集中管理 IT 资源，联合各个系统，提高资源利用率和降低成本的。

　　[8] 长句。句子结构为 This type of…allows…, and enables…；and 连接的前后两个句子中，分别有 with…和 to meet…两个状语。

　　[9] delivered…过去分词短语作定语，修饰 services；appearing…现在分词短语作状语。

　　[10] Google App Engine（GAE）是一种可以在谷歌的基础架构上运行的网络应用程序。Google App Engine 应用程序易于构建和维护，并可根据访问量和数据存储需要的增长轻松扩展。Amazon Web Services（AWS）是亚马逊提供的一种云计算服务。通过这个平台，可以将多余的资源提供给有需要的人。Microsoft Azure 是微软基于云计算的操作系统，它和 Azure Services Platform 一样，是微软"软件和服务"技术的名称。Windows Azure 的主要目标是为开发者提供一个平台，帮助开发可运行在云服务器、数据中心、Web 和 PC 上的应用程序。IBM Blue Cloud 是 IBM 推出的"蓝云"计划，它为客户带来即可使用的云计算。Salesforce 是创建于 1999 年 3 月的一家客户关系管理（CRM）软件服务提供商，总部设于美国旧金山，可提供随需应用的客户关系管理平台。

　　[11] VM（Virtual Machine）即虚拟机，指通过软件模拟的、具有完整硬件系统功能的、运行在一个完全隔离环境中的完整计算机系统。

　　[12] intranet，内联网，有关解释，见 5.2.1 节 NOTES [2]。

　　[13] 长句。中间插入 built by IBM，that 引导的是定语从句，修饰 private cloud。IBM Research Center 是 IBM 公司的研究和开发部门，是世界上最大的研究机构之一。

　　[14] 长句，句中 that 引导的定语从句修饰 accessibility；后面的 in the same way the user interface…句中 the user interface 前省略了 by which。

　　[15] REST 的原文是 Representational State Transfer，它是分布式超媒体系统（如万维网）的一种软件体系结构格式。

KEYWORDS

cloud computing	云计算
delivery	交付，投递
electricity grid	电网
utility	实用程序
end-user	最终用户，直接用户
thin client	瘦客户
data center	数据中心
manageability	可管理性

maintenance	维护
infrastructure	基础设施，基础结构
infrastructure convergence	基础会聚
public cloud	公共云
subscription	订购，订阅
accessible	可访问的，可接入的
callout	标注
architecture	体系结构，架构
VM(Virtual Machine)	虚拟机
private cloud	私有云
client	客户
hybrid cloud	混合云
RC2(Research Compute Cloud)	研究计算云
scatter	撒播，分散
pool of users	用户群
redundant	冗余的

EXERCISES

Multiple Choices

1. Cloud computing is _____.
   a. a marketing term for technologies   b. the delivery of computing
   c. a product   d. a new supplement for IT services
2. With cloud computing we can share _____.
   a. resources   b. software   c. information   d. services
3. Using cloud computing we should _____.
   a. know the physical location of the system
   b. not know the physical location of the system
   c. connect to Internet
   d. know the configuration of the system
4. Cloud computing deliveries _____.
   a. virtualized resources   b. dynamically scalable resources
   c. the form of Web-based tools   d. a delivery model for IT services
5. The applications delivered by cloud computing providers can be accessed from _____.
   a. a person who has not any of electronic devices
   b. mobile apps
   c. desktop computer
   d. Web browser
6. In cloud computing we can encounter _____.

CHAPTER 5 NEW INTERNET APPLICATIONS

   a. legacy applications   b. screen-sharing technology
   c. Web-based technologies   d. data-centers
7. One type of data center environment allows enterprises to _____.
   a. get their applications up   b. run applications slower
   c. manage applications easier   d. adjust IT resources more rapidly
8. The tremendous impact of cloud computing on business has prompted _____ in this text.
   a. British   b. federal U.S. government
   c. many agencies   d. state government in U.S.
9. A public cloud _____.
   a. can be accessed free by any user
   b. can be accessed by any user who has paid for the service
   c. can be accessed through a subscription
   d. is owned by service provider
10. Right now we can access the following public clouds _____.
    a. IBM Blue Cloud   b. Salesforce
    c. Microsoft Azure   d. Amazon Web Services
11. A commercial public cloud provides _____.
    a. an infrastructure service on a flexible price-per-use basis
    b. an infrastructure service on a fixed price-per-use basis
    c. a public accessible remote interface
    d. a selected set of business processes
12. A private cloud is _____.
    a. client owned and managed   b. server owned and managed
    c. owned by a single organization d. owned by multiple organizations
13. A private cloud can be accessed by _____.
    a. any public user   b. the owning clients
    c. their partners   d. users within their administrative domains
14. A private cloud _____.
    a. may impact the cloud standardization
    b. retain greater customization and organizational control
    c. is Internet-based
    d. is Intranet-based
15. A hybrid cloud _____.
    a. is built with both public and private clouds
    b. can be supported by private clouds
    c. provides access to clients
    d. provides access to third parties
16. In cloud computing the computing resources can be _____.

a. controlled by users themselves   b. controlled by centralized IT service
c. re-provided to consumers   d. accessed by APIs

17. Accessing cloud computing systems, we should not concern _____.
    a. the location of required devices   b. what device is used
    c. if the Internet is connected or not   d. if the used browser is online or not

18. In cloud computing the following features are improved _____.
    a. security   b. reliability   c. agility   d. maintenance

5.4 BIG DATA

Big data is a broad term for data sets so large or complex that traditional data processing applications are inadequate[1]. Challenges include analysis, capture, curation, search, sharing, storage, transfer, visualization, and information privacy[2]. The term often refers simply to the use of predictive analytics or other certain advanced methods to extract value from data, and seldom to a particular size of data set.

Analysis of data sets can find new correlations, to "spot business trends, prevent diseases, combat crime and so on." Scientists, Practitioners of Media and Advertising and Governments alike regularly encounter limitations due to large data sets in many areas. The limitations affect Internet search, finance and business informatics.

Scientists, for example, encounter limitations in e-Science work, including meteorology, genomics, connectomics, complex physics simulations, and biological and environmental research[3].

Data sets grow in size in part because they are increasingly being gathered by cheap and numerous mobile devices, aerial sensory technologies (remote sensing), software logs, cameras, microphones, radio-frequency identification (RFID) readers, and wireless sensor networks[4]. The world's technological per-capita capacity to store information has roughly doubled every 40 months since the 1980s; as of 2012, every day 2.5 exabytes (2.5×10^{18}) of data were created; the challenge for large enterprises is determining who should own big data initiatives that straddle the entire organization[5].

Figure 5-3 shows the growth of global information storage capacity.

1. Definition

Big data is a popular term used to describe the exponential growth, availability and use of information, both structured and unstructured. According to IDG, it is imperative that organizations and IT leaders focus on the ever-increasing volume, variety, and velocity of information that forms big data[6].

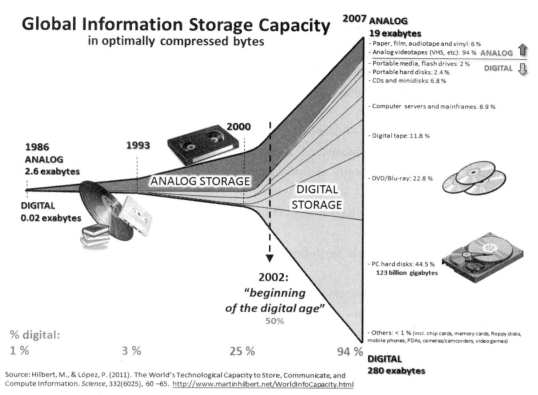

Figure 5-3   Growth and Digitization of Global Information Storage Capacity Source

(1) Volume

Many factors contribute to the increase in data volume-transaction-based data stored through the years, text data constantly streaming in from social media, increasing amounts of sensor data being collected, etc. In the past, excessive data volume created a storage issue. But with today's decreasing storage costs, other issues emerge, including how to determine relevance amidst the large volumes of data and how to create value from that is relevant.

(2) Variety

Data today comes in all types of formats—from traditional databases to hierarchical data stores created by end users and OLAP systems, to text documents, E-mail, meter-collected data, video, audio, stock ticker data and financial transactions[7]. By some estimates, 80 percent of an organization's data is not numeric! But it still must be included in analyses and decision making.

(3) Velocity

According to Gartner, velocity "means both how fast data is being produced and how fast the data must be processed to meet demand". RFID tags and smart metering are driving an increasing need to deal with torrents of data in near-real time.

2. Uses for big data

So the real issue is not that you are acquiring large amounts of data (because we are clearly already in the era of big data). It's what you do with your big data that matters[8]. The hopeful vision for big data is that organizations will be able to harness relevant data and use it to make the best decisions.

Technologies today not only support the collection and storage of large amounts of data, they provide the ability to understand and take advantage of its full value, which helps organizations run more efficiently and profitably[9]. For instance, with big data and big data analytics, it is possible to:

- Analyze millions of SKUs to determine optimal prices that maximize profit and clear inventory[10].
- Recalculate entire risk portfolios in minutes and understand future possibilities to mitigate risk.
- Quickly identify customers who matter the most.
- Generate retail coupons at the point of sale based on the customer's current and past purchases, ensuring a higher redemption[11].
- Send tailored recommendations to mobile devices at just the right time, while customers are in the right location to take advantage of offers.
- Analyze data from social media to detect new market trends in demand.
- Use clickstream analysis and data mining to detect fraudulent behavior[12].
- Determine root causes of failures, issues and defects by investigating user sessions, network logs and machine sensors.

Examples of big data:

- RFID systems generate up to 1000 times the data of conventional bar code systems.
- 10000 payment card transactions are made every second around the world.
- Walmart handles more than 1 million customer transactions an hour.
- 340 million tweets are sent per day. That's nearly 4000 tweets per second.[13]
- Facebook has more than 901 million active users generating social interaction data.
- More than 5 billion people are calling, texting, tweeting and browsing websites on mobile phones.

3. Technologies

A number of recent technology advancements are enabling organizations to make the most of big data and big data analytics:

- Cheap, abundant storage and server processing capability.
- Faster processors.
- Affordable large-memory capabilities, such as Hadoop[14].

- New storage and processing technologies designed specifically for large data volumes, including unstructured data.
- Parallel processing, clustering, MPP, virtualization, large grid environments, high connectivity and high throughputs[15].
- Cloud computing and other flexible resource allocation arrangements.

Big data technologies not only support the ability to collect large amounts of data, they provide the ability to understand it and take advantage of its value. The goal of all organizations with access to large data collections should be to harness the most relevant data and use it for optimized decision making.

It is very important to understand that not all of your data will be relevant or useful. But how can you find the data points that matter most? It is a problem that is widely acknowledged. "Most businesses have made slow progress in extracting value from big data. And some companies attempt to use traditional data management practices on big data, only to learn that the old rulers no longer apply," says Dan Briody, in the 2011 Economist Intelligence Unit's publication, "Big Data Harnessing a Game-Changing Asset." [16]

NOTES

[1] that 引导的是结果状语从句。

[2] curation，策展，即策划、筛选并展示的意思。早期定义是艺术展览活动中的构思、组织、管理工作。

[3] e-Science，由英国在 2000 年提出，是为了应对当时各学科研究领域所面临问题的空前复杂化，利用新一代网络技术（Internet）和广域分布式高性能计算环境（Grid）建立的一种全新科学研究模式，即在信息化基础设施支持下的科学研究活动。connectomics（连接组学）是近年来一系列生命科学研究中的一个分支。

[4] 长句。because 引导的是原因状语从句。有关 RFID 的内容，详见 4.3.2 节，NOTES[4]。

[5] 用两个分号隔开的长句。最后一句中 who 引导的是宾语从句。

[6] 第一个 that 引导的是主语从句，第二个 that 引导的是定语从句，修饰 information。IDG（International Data Group）即国际数据集团，是全世界最大的信息技术出版、研究、发展与风险投资公司。IDG 公司 2005 年全球营业总收入达到 26.8 亿美元。IDG 集团公司创建于 1964 年，总部设在美国波士顿。

[7] 长句。OLAP（On-Line Analytical Processing）即联机分析处理，是数据仓库系统最主要的应用，专门设计用于支持复杂的分析操作，侧重对决策人员和高层管理人员的决策支持。

[8] what 引导的是表语从句。 that 引导的是定语从句，修饰 big data。matter，此处为关系重大之意。

[9] 长句。which 引导的是非限定性定语从句。

[10] SKU(Stock Keeping Unit)，库存量单位，即库存进出计量的单位，可以以件、盒、托盘等为单位。

[11] redemption，兑换率，回收，偿还。

[12] clickstream，点击流，一个用来跟踪正在你的站点上访问的用户所到达位置细节的工具。它允许你跟踪访问你的站点的"点击流"或者"传输路径"。data mining，数据挖掘，又译为资料探勘、数据采矿。它是数据库知识发现中的一个步骤。数据挖掘一般是指从大量的数据中通过算法搜索隐藏于其中信息的过程。

[13] tweet，推文，有关它的内容详见 5.1.3 节。

[14] Hadoop，由 Apache 基金会开发的分布式系统基础架构，用户可以在不了解分布式底层细节的情况下，开发分布式程序。充分利用群集的威力进行高速运算和存储。

[15] MPP（Massively Parallel Processing）意为大规模并行处理系统，这样的系统是由许多松耦合处理单元组成的。

[16] 此处引用 Dan Briody 说的原话，前后两段分别用引号标明，中间是插入语：says Dan Briody，in…。Economist Intelligence Unit's publication，经济学人智库出版物。

KEYWORDS

big data	大数据
data set	数据集，数据传输机
capture	捕获，捕捉，截获
curation	策展
visualization	可视化，目视，显像
predictive analytics	预测分析
informatics	信息控制论，信息科学，情报学，资料学
genomics	基因学
connectomics	连接组学
biological	生物学的，生物的
remote sensing	远程传感技术
RFID(Radio Frequency Identification)	射频识别
per-capita	人均的，人均
exabyte(10^{18} byte)	艾字节
exponential	指数，指数的，幂
IDG(International Data Group)	国际数据集团
stream	流，序列
social media	社交媒体
sensor	传感器
database	数据库
hierarchical	层次的，分层的
OLAP(On-Line Analytical Processing)	联机分析处理
smart metering	智能仪表
parallel processing	并行处理

clustering　　　　　　　　　　　　　聚类，分类归并，群集，分群，分组，划分
MPP(Massively Parallel Processing)　　大规模并行处理系统
grid　　　　　　　　　　　　　　　　格栅，栅极
throughput　　　　　　　　　　　　　吞吐量，通过量，总处理能力
decision making　　　　　　　　　　　决策支撑

EXERCISES

True/False

1. _____ Big data can be used to process very complex data.
2. _____ Usually big data is directed against a particular size of data set.
3. _____ Analysis of data set can find new correlations among events.
4. _____ Scientists alike regularly encounter limitations as large data set.
5. _____ e-Science is a new idea introduced by America.
6. _____ It is important to possess the big data initiatives for large enterprises.
7. _____ The ever-increasing volume, variety, and velocity of information promote big data.
8. _____ Right now excessive data volume creates storage issue.
9. _____ On-line Analytical Processing (OLAP) is the major application of data mining technology.
10. _____ About 80% of an organization's data is numeric.
11. _____ Velocity in big data means how fast data is being produced only.
12. _____ We are clearly already in the era of big data.
13. _____ Organizations can use big data to make the best decisions.
14. _____ Stock Keeping Unit (SKU) is used to determine optimal prices.
15. _____ A point of sale should generate retail coupons only based on the customer's current purchases.
16. _____ In order to detect fraudulent behavior we can use clickstream analysis.
17. _____ We have listed six samples of big data.
18. _____ Cloud computing can help organizations to make the most of big data.
19. _____ Big data technologies only support the ability to collect large amounts of data.
20. _____ Most businesses have made fast progress in extracting value from big data.

PART III

PROGRAMMING LANGUAGES AND DATABASES

CHAPTER 6 PROGRAMMING LANGUAGES

6.1 C, C++, AND C#

C combines the best features of a structured high-level language and an assembly language—that is, it's relatively easy to code (at least compared to assembly language) and it uses computer resources efficiently [1]. Although originally designed as a system programming language (in fact, the first major program written in C was the UNIX operating system), C has proven to be a powerful and flexible language that can be used for a variety of applications [2]. It is used mostly by computer professionals to create software products.

A newer object-oriented version of C is called C++ (See Figure 6-1). C++ includes the basic features of C, making all C programs understandable to C++ compilers, but has additional features for objects, classes, and other components of an OOP. There is also a visual version of the C++ language. All in all, C++ is one of the most popular programming languages for graphical applications.

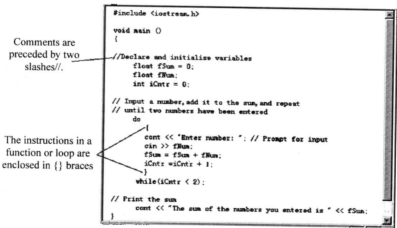

Figure 6-1 Adding-two-numbers program implemented in C++

C++ fully supports object-oriented programming, including the four pillars of object-oriented development: encapsulation, data hiding, inheritance, and polymorphism.

The property of being a self-contained unit is called encapsulation. With encapsulation, we can accomplish data hiding. Data hiding is the highly valued characteristic that an object can be used without the user knowing or caring how it works internally[3]. Just as you can use a refrigerator without knowing how the compressor works, you can use a well-designed object without knowing about its internal data members.

C++ supports the properties of encapsulation and data hiding through the creation of user-defined types, called classes [4]. Once created, a well-defined class acts as a fully encapsulated entity—it is used as a whole unit [5]. The actual inner workings of the class should be hidden. Users of a well-defined class do not need to know how the class works; they just need to know how to use it. When the engineers at Acme Motors want to build a new car, they have two choices: They can start from scratch, or they can modify an existing model. Perhaps their Star model is nearly perfect, but they'd like to add a turbocharger and a six-speed transmission. The chief engineer would prefer not to start from the ground up, but rather to say, "Let's build another Star, but let's add these additional capabilities. We'll call the new model a Quasar [6]." A Quasar is a kind of Star, but one with new features.

C++ supports the idea of reuse through inheritance. A new type, which is an extension of an existing type, can be declared. This new subclass is said to derive from the existing type and is sometimes called a derived type[7]. The Quasar is derived from the Star and thus inherits all its qualities, but can add to them as needed.

C++ supports the idea that different objects do "the right thing" through what is called function polymorphism and class polymorphism [8]. Poly means many, and morph means form. Polymorphism refers to the same name taking many forms.

While it is true that C++ is a superset of C, and that virtually any legal C program is a legal C++ program, the leap from C to C++ is very significant[9]. C++ benefited from its relationship to C for many years, as C programmers could ease into their use of C++. To really get the full benefit of C++, however, many programmers found they had to unlearn much of what they knew and learn a whole new way of conceptualizing and solving programming problems [10].

One of the major revisions of the C++ standard, C++11 (formerly known as C++0x), was approved and released on the 12 August 2011.

In 2014, C++14 (also known as C++1y) was released as a small extension to C++11, featuring mainly bug fixes and small improvements. It aims at doing what C++03 did to C++98[11]. The Draft International Standard ballot procedures completed in mid-August 2014[12].

After C++14, a major revision, informally known as C++17, is planned for 2017.

The newer version of C is C# (pronounced "C sharp"). A hybrid of C and C++, C# is Microsoft's newest programming language developed to compete directly with Sun's Java language. C# is an object-oriented programming language designed to improve productivity in the development of Web applications. The most recent version is C# 5.0, which was released on August 15, 2012.

Microsoft Visual C# is Microsoft's implementation of the C# specification, included in the Microsoft Visual Studio suite of products[13]. It is based on the ECMA/ISO specification of the C# language, which Microsoft also created[14]. While multiple implementations of the specification exist, Visual C# is by far the one most commonly used.

NOTES

[1] structured language 是指结构化程序设计语言。code 为动词，写代码，即编写程序。

[2] Although 引导的是让步状语从句，主句中有 that 引导的定语从句，修饰 language。

[3] that 引导的定语从句修饰 characteristic，without 引导的是条件状语，但此句中的 knowing or caring 为独立结构，其逻辑主语为 user。

[4] 类型（type）和类（class）在 C++中都称为类。

[5] encapsulated entity…封装的实体，被作为一个整体来使用。

[6] Quasar，类星体。

[7] derived type，派生类型。

[8] that 引导的是同位语从句，through 引导的是介词宾语从句，此处作状语。

[9] While 引导的是让步状语从句，由两个 that 引导的主语从句构成。

[10] 长句。前面 To really get，为动词不定式短语作目的状语，后面一句中的谓语 found 后面省略了由 that 引出的宾语从句，此宾语从句有两个并列成分：to unlearn…and learn。

[11] doing，动名词，用作介词 at 的宾语，what 引导的是宾语从句。

[12] The Draft International Standard，是指 C++14 的标准。

[13] Microsoft Visual Studio（简称 VS）是美国微软公司的开发工具包系列产品。VS 是一个基本完整的开发工具集，它包括整个软件生命周期中所需要的大部分工具，如 UM 工具、代码管控工具、集成开发环境（IDE）等。所生成的目标代码适用于微软支持的所有平台。

[14] ECMA/ISO，ECMA（European Computer Manufacturers Association），欧洲计算机制造联合会，是 1961 年成立的旨在建立统一的计算机操作格式标准——包括程序语言和输入输出的组织。ISO（International Standardization Organization）即国际标准化组织。

KEYWORDS

structured language	结构化语言
OOP (Object-Oriented Programming)	面向对象程序设计
compiler	编译器（程序）
class	类
Java language	一种面向对象的程序设计语言
encapsulation	包装，封装
data hiding	数据隐藏
inheritance	继承性
polymorphism	多态性
portable	可移植的

ANSI	美国国家标准学会
superset	超类
bug	故障，错误，安全漏洞
fix	修正，修正程序
specification	规范，规格，说明书，技术要求
suite	套件，程序集，一套，一组
unqualified	非限定的，不合格的
reference	引用，参考，基准，坐标，标记
context	上下文，前后关系，语言环境，场合

EXERCISES

Multiple choices

1. C language possesses best features of _____.
   a. a structured high-level language   b. coding relatively easy
   c. using computer resources efficiently   d. an assembly language
2. C++ include the following pillars: _____.
   a. encapsulation   b. data hiding
   c. inheritance   d. polymorphism
3. Encapsulation is _____.
   a. derived from data hiding   b. a self-contained unit
   c. used to accomplish data hiding   d. supported by C++
4. Classes in C++ _____.
   a. are user-defined types
   b. can be used to support the properties of encapsulation and data hiding
   c. act as a fully encapsulated entity
   d. like a new car started from scratch
5. A new type _____.
   a. is a subclass
   b. can be derived from an existing type
   c. can't be called a derived type
   d. inherits all features of existing type and has additional features
6. Polymorphism in C++_____.
   a. means the same name taking many forms
   b. supports different objects do the right thing
   c. has function polymorphism
   d. has class polymorphism
7. C++ is _____.
   a. object-oriented   b. procedure-oriented

    c. extended from C   d. a superset of C
8. The actual inner workings of the class _____.
    a. should not be hidden   b. should be hidden
    c. like a "black box"   d. don't like a "black box"
9. C# is _____.
    a. a hybrid of C and C++
    b. developed to compete indirectly with Sun's Java language
    c. designed to improve productivity for developing Web applications
    d. the newer version of C
10. Visual C# is _____.
    a. MS's implementation of the C# specification
    b. based on the ECMA/ISO specification of the C# language
    c. by far the one most commonly used
    d. included in the MS Visual Studio suite of products

6.2   JAVA

1. Overview of Java

Java is a programming language originally developed by James Gosling at Sun Microsystems (now part of Oracle Corporation) and released in 1995 as a core component of Sun Microsystems' Java platform[1]. The language derives much of its syntax from C and C++ but has a simpler object model and fewer low-level facilities. Java applications are typically compiled to bytecode (class file) that can run on any Java Virtual Machine (JVM) regardless of computer architecture[2]. Java is a general-purpose, concurrent, class-based, object-oriented language that is specifically designed to have as few implementation dependencies as possible[3]. It is intended to let application developers "write once, run anywhere." (WORA), meaning that compiled Java code can run on all platforms that support Java without the need for recompilation.Java is one of the most popular programming languages in use, particularly for client-server web applications, with a reported 9 million developers.

2. Goals

There were five primary goals in the creation of the Java language:
- It should be "simple, object-oriented and familiar"
- It should be "robust and secure"
- It should be "architecture-neutral and portable"
- It should execute with "high performance"
- It should be "interpreted, threaded, and dynamic"

3. Java Platform

One characteristic of Java is portability, which means that computer programs written in the Java language must run similarly on any hardware/operating-system platform[4]. This is achieved by compiling the Java language code to an intermediate representation called Java bytecode, instead of directly to platform-specific machine code. Java bytecode instructions are analogous to machine code, but are intended to be interpreted by a virtual machine (VM) written specifically for the host hardware. End-users commonly use a Java Runtime Environment (JRE) installed on their own machine for standalone Java applications, or in a Web browser for Java applets [5].

Standardized libraries provide a generic way to access host-specific features such as graphics, threading, and networking.

A major benefit of using bytecode is porting. However, the overhead of interpretation means that interpreted programs almost always run more slowly than programs compiled to native executables world[6]. Just-in-Time compilers were introduced from an early stage that compile bytecodes to machine code during runtime[7].

4. Java applet

A Java applet is an applet delivered to users in the form of Java bytecode. Java applets can run in a Web browser using a Java Virtual Machine (JVM), or in Sun's AppletViewer, a stand-alone tool for testing applets [8].

Java applets run at speeds comparable to, but generally slower than, other compiled languages such as C++, but until approximately 2011 many times faster than JavaScript [9]. In addition they can use 3D hardware acceleration that is available from Java. This makes applets well suited for non-trivial, computation intensive visualizations. When browsers have gained support for native hardware accelerated graphics in the form of Canvas and WebGL, as well as Just in Time compiled JavaScript, the speed difference has become less noticeable [10].

Since Java's bytecode is cross-platform or platform independent, Java applets can be executed by browsers for many platforms, including Microsoft Windows, UNIX, Mac OS and Linux. It is also trivial to run a Java applet as an application with very little extra code. This has the advantage of running a Java applet in offline mode without the need for any Internet browser software and also directly from the integrated development environment (IDE)[11].

NOTES

[1] as 引导的为方式状语。Sun Microsystems 曾为著名的微型计算机系统公司，以生产工作站出名，后并入 Oracle（俗称甲骨文）公司。

[2] that 引导的定语从句，修饰 bytecode。bytecode, Java 字节码, 是一种中间语言，通常它把 Java 源代码编译成字节码，也可以把其他语言，如 Ada 语言的源代码编译成字

节码。JVM（Java Virtual Machine）是在虚拟和非虚拟硬件上，在标准操作系统上实现的软件，它提供一个可运行 Java 字节码的环境。

[3] implementation dependencies，实现依赖，是说 union 里的数据，如果你存储时用一种数据类型，然后用另一种数据类型去访问，那么访问得到的结果是未知的。在 C++ 规范中没有规定应该会有怎么样的结果，而依赖于编译器的实现，在不同的编译器中，不同的优化条件可以有不同的返回结果，这就叫实现依赖。

[4] which 引导的是非限定性定语从句。

[5] 长句，JRE（Java Runtime Environment）（包括 JavaPlug-in）是 Sun 公司的产品，包括两部分：Java Runtime Environment 和 JavaPlug-in。JRE 是可以在其上运行、测试和传输应用程序的 Java 平台。它包括 Java 虚拟机、Java 平台核心类和支持文件。它不包含开发工具——编译器、调试器和其他工具。JRE 需要辅助软件 JavaPlug-in，以便在浏览器中运行 applet。

[6] that 引导的是宾语从句，从句中有 more…than 结构。

[7] that 引导的定语从句，修饰 compilers。

[8] 由 or 连接的并列句，句子主语为 Java applets；后一句中省略 can run，此处的 applets 是指另外的小应用程序；Sun 公司的 AppletViewer（小应用程序观察器）是 JDK（Java Development Kit）自带的工具，用于快速测试小应用程序。

[9] JavaScript 的解释见 6.3 节。

[10] When 引导的是时间状语从句。WebGL(Web-based Graphics Library)是一种扩展了 JavaScript 程序设计语言功能的软件库，它可在任何兼容的 Web 浏览器内，制作交互式 3D 图形。Canvas 是一种 HTML 元件，它含有 WebGL 文本。

[11] 并列句，主语为 This，and 后面省略了 runs。IDE（Integrated Development Environment）即集成开发环境，又称集成设计环境、集成调试环境或交互式开发环境，它为程序员提供了很多复杂功能。

KEYWORDS

Java	一种程序设计语言
Java platform	Java 平台
object model	对象模型
bytecode	字节代码
class file	类文件
JVM(Java Virtual Machine)	Java 虚拟机
concurrent	并发（行）的
object-oriented	面向对象的
portable	可移植的
interpret	解释
JRE(Java Runtime Environment)	Java 运行时环境
library	（程序）库

overhead	开销
compiler	编译程序
recompilation	再次编译
client-server	客户-服务器（方式）
Java applet	Java 小应用程序
visualization	可视化
cross-platform	交叉平台
IDE(Integrated Development Environment)	集成开发环境

EXERCISES

Fill in the blanks with using appropriate words of terms found behind this exercises.

1. Java was originally developed at _____.
2. Java derives much of its syntax from _____.
3. _____ are typically compiled to bytecode.
4. Java is a _____ language.
5. There were _____ in creation of the Java language.
6. _____ should be interpreted.
7. _____ WORA means.
8. _____ is an intermediate representation.
9. _____ provide a generic way to access host-specific features.
10. A major benefit of using bytecode is _____.
11. End-users commonly use a _____ installed on their own machine for standalone Java applications.
12. Just-in-Time compilers can _____ bytecode to machine code during runtime.
13. Running Java applet we can use _____.
14. Java applets run generally slower than _____.
15. _____ runs many times slower than Java applets.
16. _____ makes applets well suited for computation intensive visualization.
17. Java's bytecode is _____.
18. Java applets can be executed for _____.

    a. five primary goals

    b. Java Runtime Environment

    c. Sun Microsystem

    d. JavaScript

    e. Unix, Mac OS, and Linux

    f. C and C++

    g. Standardized libraries

    h. 3D hardware acceleration

i. Java Virtual Machine

j. write once, run anywhere

k. concurrent, class-based and object-oriented

l. porting

m. cross-platform

n. Java applications

o. Java bytecode

p. Java

q. compiled languages

r. compile

6.3 MARKUP AND SCRIPTING LANGUAGES

There are languages other than programming languages that are used in conjunction with application development. The majority of these are Web related, as discussed in the next few sections.

1. HTML and Other Markup Languages

(1) HTML

Most Web pages today are written in a markup language. Markup languages are designed to make it possible to transmit documents over a network using minimal line capacity[1]. Instead of sending exact specifications regarding the appearance of a Web page, markup languages define the structure and layout of a Web page by using a variety of tags[2]. The most common markup language for Web pages is HTML(Hypertext Markup Language). HTML uses HTML tags.

When a Web page is created—using either a word processor, text editor, or a special Web site development program—HTML tags are inserted in the appropriate locations within the Web page's text. Some tags are used along; others are used in pairs. For example, <b> turns bolding on for the text that follows up until a tag </b> is reached, so the following HTML statement

    <b>This text is bolded. </b>

would produce the following when viewed with most Web browsers.

This text is bolded

A Web page and its corresponding HTML code are shown in Figure 6-2, with some common HTML tags.

HTML5 is a core technology markup language of the Internet used for structuring and presenting content for the World Wide Web. As of October 2014, this is the final and complete fifth revision of the HTML standard of the World Wide Web Consortium (W3C)[3]. The previous version, HTML 4, was standardized in 1997.

Figure 6-2   HTML

Its core aims have been to improve the language with support for the latest multimedia while keeping it easily readable by humans and consistently understood by computers and devices (web browsers, parsers, etc.)[4]. HTML5 is intended to subsume not only HTML4, but also XHTML1 and DOM Level 2 HTML[5].

In particular, HTML5 adds many new syntactic features. These include the new <video>, <audio> and <canvas> elements, as well as the integration of scalable vector graphics (SVG) content (replacing generic <object> tags), and MathML for mathematical formulas[6]. These features are designed to make it easy to include and handle multimedia and graphical content on the web without having to resort to proprietary plugins and APIs. Other page structuring new elements, such as <main>, <section>, <article>, <header>, <footer>, <aside>, <nav> and <figure>, are designed to enrich the semantic content of documents. New attributes have been introduced for the same purpose, while some elements and attributes have been removed[7]. Some elements, such as <a>, <cite> and <menu> have been changed, redefined or standardized. The APIs and Document Object Model (DOM) are no longer afterthoughts, but are fundamental parts of the HTML5 specification. HTML5 also defines in some detail the required processing for invalid documents so that syntax errors will be treated uniformly by all conforming browsers and other user agents.

(2) XML

Extensible Markup Language (XML) is a markup language that defines a set of rules for encoding documents in a format which is both human-readable and machine-readable[8]. It is defined by the W3C's XML 1.0 Specification and by several other related specifications, all of which are free open standards.

The design goals of XML emphasize simplicity, generality and usability across the Internet. It is a textual data format with strong support via Unicode for different human languages[9]. Although the design of XML focuses on documents, it is widely used for the representation of arbitrary data structures, such as those used in web services.

Several schema systems exist to aid in the definition of XML-based languages, while many application programming interfaces (APIs) have been developed to aid the processing of XML data.

(3) .NET

Very closely interrelated with XML is Microsoft's .NET strategy to increase the convergence of personal computing with the Web[10]. In a nutshell, .NET is Microsoft's platform to implement XML-based Web services. These services allow applications to communicate and share data over the Internet, regardless of the operating system or programming language being used.

2. Scripting Languages

(1) JavaScript

HTML is principally designed for laying out Web pages that have on moving elements, much as a desktop-publishing program is designed for laying out printed pages. Thus, HTML has minimal tools to create Web pages that change as the user looks at them or to enable users to interact with Web pages on their screens, other than some capabilities with DHTML and recent HTML enhancements[11]. If you want to develop pages with a great deal of dynamic content, scripting languages, such as JavaScript, may be appropriate. Such languages enable you to build program instructions, or scripts, directly into a Web page's code to add dynamic content. For example, JavaScript is commonly used to display submenus or new images when a menu item is pointed to (See Figure 6-3).

JavaScript was originally developed by Netscape to enable Web authors to implement interactive Web sites. Although it shares many of the features and structures of the full Java language, it was developed independently. When using JavaScript, it is important to realize that not all scripting commands work with all browsers[12]. Because of this, make sure that the important features you add to your site with JavaScript are not browser specific.

(2) PHP

PHP is a server-side scripting language designed for web development but also used as a general-purpose programming language. As of January 2013, PHP was installed on more than 240 million websites (39% of those sampled) and 2.1 million web servers.

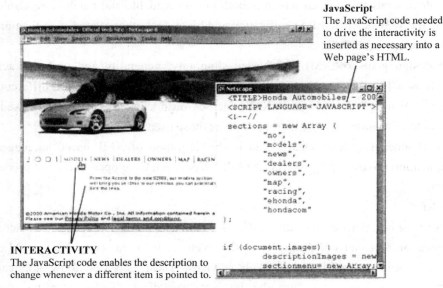

Figure 6-3　JavaScript

PHP code can be simply mixed with HTML code, or it can be used in combination with various templating engines and web frameworks[13]. PHP code is usually processed by a PHP interpreter, which is usually implemented as a web server's native module or a Common Gateway Interface (CGI) executable[14]. After the PHP code is interpreted and executed, the web server sends resulting output to its client, usually in form of a part of the generated web page; for example, PHP code can generate a web page's HTML code, an image, or some other data. PHP has also evolved to include a command-line interface (CLI) capability and can be used in standalone graphical applications[15].

The canonical PHP interpreter, powered by the Zend Engine, is free software released under the PHP License[16]. PHP has been widely ported and can be deployed on most web servers on almost every operating system and platform, free of charge.

Despite its popularity, no written specification or standard existed for the PHP language until 2014, the canonical PHP interpreter is as a de facto standard[17]. Since 2014, there is ongoing work on creating a formal PHP specification.

As of 2014, work is underway on a new major PHP version named PHP 7.

NOTES

[1] minimal line capacity，此处指用标记语言写的网页简明，在网上传输时占用的带

宽少。

[2] 长句，Instead of…引导的介词短语作状语，主句为 markup languages…。

[3] W3C（World Wide Web Consortium）即万维网联盟，创建于 1994 年，是 Web 技术领域最具权威和影响力的国际中立性技术标准机构。到目前为止，W3C 已发布了 200 多项影响深远的 Web 技术标准及实施指南，如广为业界采用的超文本标记语言（标准通用标记语言下的一个应用）、可扩展标记语言（标准通用标记语言下的一个子集），以及帮助残障人士有效获得 Web 内容的信息无障碍指南（WCAG）等，有效促进了 Web 技术的互相兼容，对互联网技术的发展和应用起到了基础性和根本性的支撑作用。

[4] 长句。while 连接的 keeping it…为分词短语，作伴随状语。

[5] XHTML1 是 XML 风格的 HTML 4.01。但还没等 XHTML 兴起，它的地位就被 HTML5 取代了。DOM Level2 HTML，其中 HTML DOM（Document Object Model）定义了一套标准的针对 HTML 文档的对象。DOM level1 模型：将 HTML 文档封装成了对象；DOM level2 模型：在 level1 的基础上，加入了名称空间的功能。

[6] MathML，数学置标语言，是一种基于 XML 的标准，用来在互联网上书写数学符号和公式的置标语言。

[7] while 连接了两个并列句。

[8] that 引导的定语从句中又有 which 引导的定语从句。

[9] Unicode（统一码、万国码、单一码）是一种在计算机上使用的字符编码。Unicode 是为了解决传统的字符编码方案的局限而产生的，它为每种语言中的每个字符设定了统一并且唯一的二进制编码，以满足跨语言、跨平台进行文本转换、处理的要求。

[10] 过去分词 interrelated 作表语，此句强调表语，为倒装句。

[11] 长句，主句中 tools 有两个并列的动词不定式作定语，即 to create…or to enable…。而第一个定语中，又有 that 引导的定语从句修饰 pages。DHTML 为动态超文本标记语言，用于在 Web 页面中增加诸如移动、增长、收缩、隐蔽或出现等动态特性的标记语言。

[12] When 引导的是时间状语，it 是先行代词，为形式主语，实际的主语是 to realize…，that 引导的是宾语从句。

[13] templating engine，模板引擎（这里特指用于 Web 开发的模板引擎），是为了使用户界面与业务数据（内容）分离而产生的，它可以生成特定格式的文档，用于网站的模板引擎会生成一个标准的 HTML 文档。

[14] CGI（Common Gateway Interface）即通用网关接口，是 WWW 技术中最重要的技术之一，有着不可替代的重要地位。CGI 是外部应用程序（CGI 程序）与 Web 服务器之间的接口标准，是在 CGI 程序和 Web 服务器之间传递信息的规程。

[15] CLI（Command-Line Interface）即命令行接口，是指可在用户提示符下输入可执行指令的界面，它通常不支持鼠标，用户通过键盘输入指令，计算机接收到指令后，予以执行。

[16] Zend Engine 是 Zend 引擎，是 PHP 实现的核心，提供了语言实现上的基础。Zend 引擎最主要的特性就是把 PHP 的边解释边执行的运行方式改为先预编译（Compile），然后再执行（Execute）。

[17] 长句。Despite 引导的是让步状语从句。

KEYWORDS

markup language	标记（置标）语言
scripting language	脚本[描述]语言，过程[编制]语言
HTML (Hypertext Markup Language)	超文本标记语言
line capacity	线路容量
tag	标记
word processor	字处理器（软件）
text editor	文本编辑器（软件）
W3C(World Wide Web Consortium)	万维网联盟
browser	浏览器
parser	语法分析程序
subsume	包含，包括
syntactic	句法的，句法上的
SVG(Scalable Vector Graphics)	可缩放的矢量图形
plugin	插件
API(Application Program Interface)	应用程序接口
XML(eXtensible Markup Language)	可扩展的标记语言
JavaScript	Java 过程（程序）语言，基于对象的脚本语言
DHTML (Dynamic HTML)	动态超文本标记语言
attribute	性质，特性，本性，属性，象征
human-readable	人可读的
machine-readable	机器可读的
interpreter	解释程序

EXERCISES

True/False

1. _____ Markup languages are common programming languages.
2. _____ Markup language is used to write Web pages.
3. _____ We must send exact specifications regarding the appearance of a Web page written by using a markup language.
4. _____ HTML uses a variety of tags.
5. _____ We must use tags of HTML in pairs.
6. _____ HTML5 is a core technology markup language of the Internet.
7. _____ If we want to increase the convergence of personal computing with the Web, we should use .NET platform.

CHAPTER 6 PROGRAMMING LANGUAGES

8. _____ The services provided by .NET platform allow applications to communicate and share data over the Internet.

9. _____ A scripting language is used for developing Web pages with a great deal of dynamic content.

10. _____ It is important to realize that all scripting commands of JavaScript work with all browsers.

11. _____ HTML5 is intended to subsume only HTML4.

12. _____ HTML5 adds many syntactic features.

13. _____ The APIs and DOM in HTML5 are afterthoughts.

14. _____ XML has a free open standard.

15. _____ XML emphasize simplicity, generality and usability across the Internet.

16. _____ PHP can be used in combination with various template engines and web frameworks.

17. _____ PHP is a client-side scripting language as well.

18. _____ PHP code is usually processed by a PHP compiler.

19. _____ After the PHP code is executed, the web server sends resulting output to its client.

20. _____ PHP has been widely ported and can be free of charge.

CHAPTER 7　DATABASE

7.1　DATABASE CONCEPTS

People often need to retrieve large amounts of data rapidly. An airline agent on the phone to a client may need to search quickly to find the lowest-cost flight from Atlanta to Toronto. The registrar of a university may have to scan student records swiftly to find all students who will graduate in June with a grade-point average of 3.5 or higher[1]. A clerk in a video store may need to determine if a particular movie is available for rental. The type of software used for such tasks is a database management system. Computerized database management systems are rapidly replacing the paper-based filing systems that people have had to wade through in the past to find the information their jobs require[2]. The basic features and concepts of PC-based relational database software are discussed next, using Microsoft Access as an example when needed.

1. What Is a Database Program

A database is a collection of data that is stored and organized in a manner enabling information to be retrieved as needed. A database management system (DBMS)—also sometimes called just database software—enables the creation of a database on a computer and provides easy access to data stored within it.

Although not all databases are organized identically, most PC-based databases are organized into fields, records, and files. A field is a single type of data to be stored in a database, such as a person's name or a person's telephone number. A record is a collection of related fields—for example, the ID number, name, address, and major of Phyllis Hoffman (See Figure 7-1). A file—frequently called a table in PC databases—is a collection of related records (such as all student address data, all student grade data, or all student schedule data). The resulting set of related files or tables (such as all student data) comprises the database.

The type of database software found on most PCs is the relational database management system.

2. Creating a Database

A database can contain a variety of objects (See Figure 7-2). The object created first in a new database is the table, then other objects can be created to be used in conjunction with that table as needed.

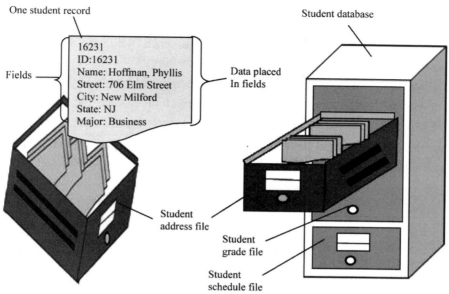

Figure 7-1 Fields, records, files, and databases. Fields, records, and files (tables) organize the data that are to be part of a database

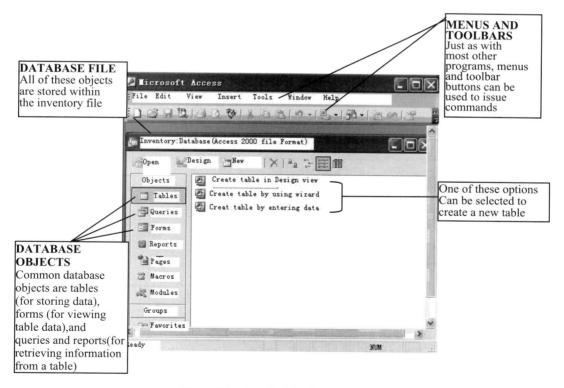

Figure 7-2 A typical database program

When creating a database, the number of tables to be included in the database should be determined. Then the data items to be stored in each table can be identified so that the appropriate field characteristics can be used[3]. For each field, the following should be determined:

① Field name (an identifying name unique within the table)

② Type of data to be contained in the field (text, numbers, date, etc.)

③ Field size (how many characters will be needed to store the data)

Once these specifications have been determined, the structure of each table containing the field specifications (See Figure7-3) can be created.

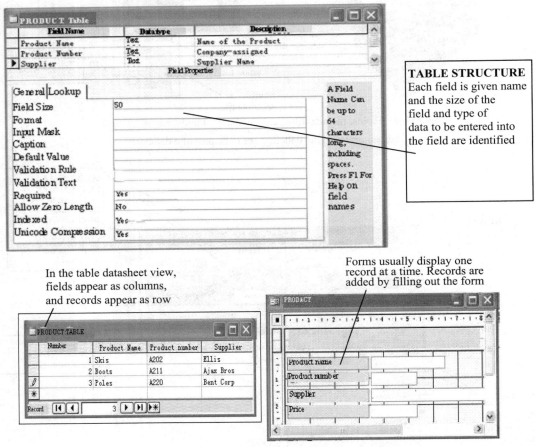

Figure 7-3　Creating a database

After the table structure has been created, data may be entered into the table. Data entry can be performed in the regular table view—sometimes called the datasheet view, since the table looks similar to a spreadsheet—or a form can be created and used[4]. A form allows you to view or edit table content in a more formal manner—usually just working with one record at a time, instead of a full page of records, as in the datasheet view. Figure 7-3 illustrates entering data using both methods once the table structure has been created.

3. Modifying a Database

Once a database table has been created, it may need to be modified. Changes may be made to the table structure or to the data located in the table as needed.

(1) Modifying the Table Structure

The table structure needs to be modified only when there are changes to the field properties. For example, a field may need to be widened to accommodate a name that is longer than anticipated, the wrong field type may have been initially selected, or a new field may need to be added.

(2) Editing, Adding, and Deleting Records

To make changes to the actual data in a table, the table must first be opened (either using the table datasheet view or a form), and then the necessary changes can be made. To move to a particular record to edit its contents, either the arrows and other keyboard directional keys or the record buttons located at the bottom of the window, can be used (refer again to Figure 7-3)[5]. Since records are typically added to the end of a table, there is usually a New Record button in the group of record buttons that automatically moves you to a blank record at the end of the table[6].

To delete a record, either the Delete key on the keyboard or some type of Delete Record option on the menu bar is used.

4. Queries and Reports

To retrieve information from a database, queries and reports are used. A query is a question, or, in database terms, a request for specific information from the database. A query takes your instructions about what information you want to find and displays just that requested information. A report is a more formal printout of a table or query result.

NOTES

[1] who 引导的定语从句，修饰 students；grade-point，美国学校学生的学习成绩的统计方法，即学绩点。

[2] 长句，句中有两个定语从句，一个是 that 引导的定语从句，修饰 filing systems；另一个是 their jobs require，省略了 that。

[3] so that 引导的是目的状语从句。

[4] 破折号中间的句子是同位语；datasheet，数据单，是一种总结产品、机器、部件、材料、子系统和软件的文档。

[5] To move…目的状语；主句结构为 either…or…can be used。

[6] Since…原因状语从句，主句中 that 引导的定语从句，修饰 New Record button。

KEYWORDS

database	数据库
DBMS (Data Base Management System)	数据库管理系统
relational database	关系型数据库
field	字段
record	记录
file	文件
ID (Identification)	标识，识别，身份证
table	表
data item	数据项
table view	表格视图
data sheet	数据单，数据表
form	表格，表单
query	查询
directional key	定向键
blank record	空记录
menu bar	菜单栏

EXERCISES

Multiple Choices

1. Database management systems can be used to search _____.
   a. the lowest-cost flight tickets　　b. student records
   c. video stock　　d. another useful data
2. MS Access is _____.
   a. PC-based　　b. a DBMS
   c. a hierarchical database　　d. a relational database
3. DBMS is _____.
   a. a database software
   b. an abbreviation for database management system
   c. used to create database
   d. used to manage database
4. Most PC-based databases are organized into _____.
   a. notes　　b. fields　　c. records　　d. files
5. The following can be used as a field _____.
   a. user's name　　b. user's telephone number
   c. product name　　d. product supplier
6. A record in a database can record the _____ of Mr. Clinton.
   a. major　　b. birth day　　c. ID number　　d. address

7. File in a database _____.
   a. is a collection of related records
   b. is called a table in PC database
   c. likes a record
   d. can be used to build a database
8. Objects in a database _____.
   a. have a variety of types
   b. can conjunct with the first created object
   c. can be tables for storing data
   d. can be formed for viewing table data
9. We should determine following specifications for each field _____.
   a. database size
   b. field size
   c. type of data to be contained in the field
   d. field name
10. Data entry can be performed in _____.
    a. a form
    b. the regular table view
    c. a spreadsheet
    d. datasheet view
11. Modifying database means _____.
    a. changing table structure
    b. changing data located in the table
    c. changing field properties
    d. adding new fields
12. To move to a particular record to edit its contents, we can use _____.
    a. a field
    b. the arrows
    c. keyboard directional keys
    d. record buttons located at the bottom of the window

7.2 THE WEB AND DATABASES

Databases are extremely common on the World Wide Web. Virtually all companies that offer products or corporate information, online ordering, or similar activities through a Web site use a database. The most common applications involve client-server database transactions, where the user's browser is the client software[1]. The use of peer-to-peer information exchange, however, is increasing[2].

1. Examples of Web Databases in Use

There are scores of examples of how databases can be used on the Web. Databases facilitate information retrieval and processing, as well as allow more interactive, dynamic content[3]. Following these sections is a discussion and example of how a Web database might work and a brief look at other Web-database-related issues.

(1) Information Retrieval

By their very nature, databases lend themselves to information retrieval on the Web, which is, in essence, a huge storehouse of data waiting to be retrieved. Data is stored in the database, and Web site visitors can request and view it (See Figure 7-4).

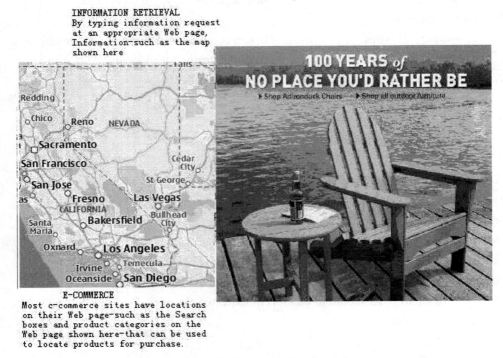

Figure 7-4　Information retrieval and e-commerce

(2) E-Commerce and E-Business

Another widely used database application on the Web is to support and facilitate e-commerce. Catalog information, pricing, customer information, shopping cart contents, and more can be stored in a database to be retrieved on demand using an appropriate script or program to link the database with the Web site [4]. (Refer again to Figure7-4).

(3) Dynamic Web Pages

Static Web pages display the same information for everyone, every time the page is displayed, until the Web page file is modified. In contrast, the appearance and content of dynamic Web pages change based on a user's input. This input can be based on selections specified on a form located on the page or controlled by some other aspect, such as a Java applet, ActiveX control, or the activities that the user has already performed on the site, such as clicking a displayed ad or a product's hyperlink [5].

2. How Web Databases Work

(1) An Example about Database and the Web Work Together

To further illustrate more about how databases and the Web can work together, let's look at an example.

The request to retrieve information from or store data into a Web database is usually initiated by the user. Filling out a Web page form, selecting an option from a menu displayed on a Web page, or clicking an onscreen are common ways database requests are made. The request is received by the Web server, which then converts the request into a database query and passes it on to the database server with the help of intermediary software called middleware. The database server retrieves the appropriate information and returns it to the Web server (again, via middleware) where it is displayed on the user's screen as a Web page. These steps are illustrated in Figure 7-5.

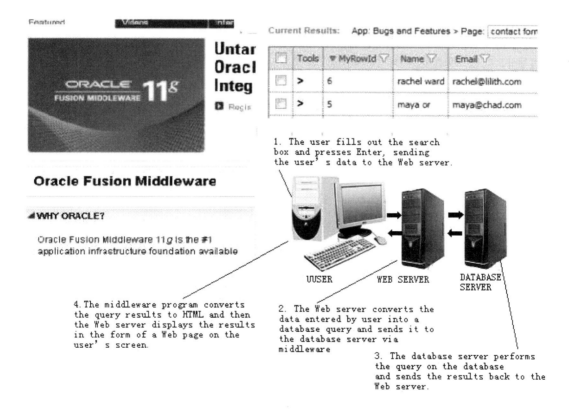

Figure 7-5　A Web database in action

(2) Middleware

Software that connects two otherwise separate applications—such as a Web server and a database management system, as in Figure7-5—is referred to as middleware. The most common types of middleware used to interface between a database and a Web page are CGI and API scripts. A newer scripting language becoming increasingly more popular is PHP and ASP [6].

NOTES

[1] where 引导的是非限定性定语从句。

[2] 与客户–服务器方式不同，peer-to-peer 对等方式是，交换信息的双方不分客户和服务器。

[3] 本句的两个谓语是 facilitate 和 allow，在后一句中 dynamic content 后面省略了 retrieval and processing。

[4] 长句。被动式动词不定式 to be retrieved…为定语修饰 database，不定式 to link 也作为定语，修饰 script or program。

[5] 长句，主句为被动式并列句：This input can be based on…or controlled by…，而后面的句子中又由 or 连接两个并列成分，且都由 such as 引入同位语。Java applet，是以 Java 字节码形式提供给用户的小应用程序，详见 6.2 节。

[6] PHP（Personal Home Page）原来的意思是个人家庭主页，但这个产品很快超出其名字的含义。现在这个缩写词的正式含义是：超文本处理程序，详见 6.3 节。ASP，动态服务器主页，是开发 Web 应用程序的、开放的、免编译的开发应用程序环境，用它可以建立强大的、综合了超文本标记语言（HTML）、脚本以及微软的 ActiveX 技术，以提供动态网站的、基于 Web 的分布式应用程序。

KEYWORDS

online ordering	在线订货
peer-to-peer	对等的
retrieval	检索
script	脚本（文件），稿本，过程
dynamic web page	动态网页
static web page	静态网页
form	表单，表格，窗体
database query	数据库查询
middle ware	中间件
CGI (Common Gateway Interface)	公用网关接口
API (Application Program Interface)	应用程序接口
PHP (Personal Home Page)	个人家庭主页
ASP (Active Server Page)	现用服务器页面，动态服务器主页

EXERCISES

True/False

1. _____ All companies offering on line services use the database via Web sites.

2. _____ There are two types of Web-based database applications, they are client-server mode and peer-to-peer mode.

3. _____ A Web site visitor can require to access a database for retrieving information.

4. _____ To link a database with a Web site we should use an appropriate script.
5. _____ The content of static Web page can be changed by a user's command.
6. _____ We can use Java applet or ActiveX control to change the content of a Web page.
7. _____ There are four common ways to require database.
8. _____ The request of retrieving information is launched by the Web server.
9. _____ In an action of Web database, the database server converts the user's request into a database query.
10. _____ Software that connects a Web server and a DBMS is called middleware.

PART IV

APPLICATION SOFTWARE

CHAPTER 8
OFFICE AUTOMATION SOFTWARE

8.1 THE BASICS OF OFFICE AUTOMATION SOFTWARE

When you use any type of application software program, such as a word processor, to type a letter or a tax preparation program to prepare your taxes, there are some basic concepts and functions you need to be familiar with [1]. These include common document-handling tasks, the concept of the software suite, ownership rights for the software you use, and how to get help while you work with the program. These topics and some features are discussed in the next few sections.

1. Basic Concepts of Office Automation Software

(1) Document-Handling Operations [2]

While some document-handling operations are specific for a particular application program, some—such as the concept of opening a document, saving it, and printing it—are fairly universal [3]. A few of the most common document-handling operations are described in Figure 8-1, with examples of the icons used to perform the operations in Windows' applications.

Start a new document		Allows you to create a new blank document, or possibly create a new document from a predefined template. The document will be stored just in RAM until it is saved onto a disk
Open a document		Opens a previously saved document from a disk, usually for editing or printing. Any changes made to the document will be stored just in RAM until the document is saved back onto the disk
Save a document		Saves the current version of the document to a disk
Print a document		Prints the current version of the document onto paper. Some programs give you a variety of print options, such as to print the entire document or just specified pages
Close a document		Removes the document from RAM. Any changes made to the document are lost if the document wasn't saved to disk before it was closed

Figure 8-1　Common document handling tasks

In general, the commands to perform these operations are the same or very similar in all GUI (Graphical User Interface) programs. Because almost everyone needs fast access to document-handling commands, these operations are usually located on easy-to-reach menus or toolbars.

(2) Software Suites

Most office-oriented programs, such as word processors and spreadsheets, are sold bundled together with other related application software in a software suite. The dominant leader in suite sales for office application is Microsoft Office. The high-end edition of this package bundles Word (for word processing), Excel (for spreadsheet work), PowerPoint (for presentation graphics), Access (for database management), together with several other programs such as FrontPage (for Web site development).

One of the biggest advantages to using a software suite is being able to transport or share documents or parts of documents from one program to another. For instance, let's say you are writing a letter in your word processing program, and you want to insert a spreadsheet table. You can launch (start) your spreadsheet program, locate the particular table you want in a stored worksheet, and then copy and paste the table into your letter—all without ever closing the word processing program (See Figure 8-2).

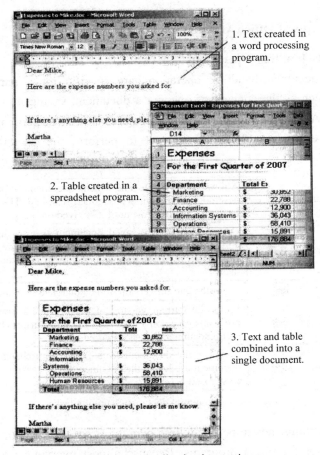

Figure 8-2 Application integration

(3) Online Help

Most people run into problems or need some help with a feature as they work with a software program. To provide help without forcing you to leave your computer screen, most application programs have an online help feature. Programs employ a variety of tools to provide online assistance. Some of the possible configurations are illustrated in Figure 8-3.

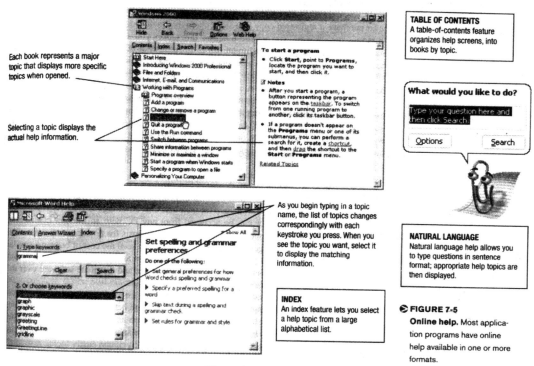

Figure 8-3　Online help

(4) Ownership and Distribution Rights

Sensitive questions sometimes arise about ownership and user rights regarding software products. Typically, a software maker or publisher develops a program, secures a copyright on it, and then retains ownership of all rights to that program[4]. The publisher then dictates who can use, copy, or distribute the program. The various classes of ownership and allowable use are discussed next.

① Proprietary Software

Many of the systems software and application program used today are proprietary software. This means that someone owns the rights to the program, and the owner expects users to buy their own copies.

② Shareware

Some software is available as shareware. While you don't have to pay to install and try out shareware, most shareware specifies that you need to pay to continue to use the software after the trial period—often one month—expires.

③ Freeware

Freeware, or public-domain software, refers to programs that you can use and share with other free of charge.

2. Common Features of Office Automation Software

A user interface is the portion of the application that you work with. Most applications use a GUI that displays graphical elements called icons to represent familiar objects and a mouse [5]. The mouse controls a pointer on the screen that is used to select items such as icons. Another feature is the use of windows to display information. A window is simply a rectangular area that can contain a document, program, or message. (Do not confuse the term window with the various versions of Microsoft's Windows operating systems, which are programs.) More than one window can be opened and displayed on the computer screen at one time.

Most software programs including Microsoft Office 2010, have menus, dialog boxes, toolbars, and buttons. (See Figure 8-4). Menus present commands that are typically displayed in a menu bar at the top of the screen. When one of the menu items is selected, an additional list of menu options or a dialog box that provides additional information and requests user input may appear[6]. Toolbars typically are below the menu bar. They contain small graphic elements called buttons that provide shortcuts for quick access to commonly used commands.

Figure 8-4 Microsoft Office Word 2007

The newest Office version, Microsoft Office 2010, has already a redesigned interface that is intended to make it easier for users to find and use all the features of an application[7]. This new design introduces ribbons, contextual tabs, galleries, and more [8].

（1）Ribbons replace menus and toolbars by organizing commonly used commands into a set of tabs. These tabs display command buttons that are the most relevant to the tasks being performed by the user.

（2）Contextual tabs are tabs that appear automatically. These tabs only appear when they are needed and anticipate the next operations to be pre-formed by the user.

（3）Galleries simplify the process of making a selection from a list of alternatives[9]. This is accomplished by replacing many dialog boxes with visual presentations of potential results.

NOTES

[1] 长句。主句为 there are…。When 引导的是时间状语从句。such as 引出的为两个并列的同位语：a word processor，…or a tax preparation…。

[2] document，文档、文件，与 file 同义，但前者多指文档，后者多指文件，本文则将其译为文档较为合适。

[3] While 引导的是让步状语从句。

[4] secure，动词，意为"确保"，此处是指在所开发的程序上保护版权。

[5] that 引导的定语从句修饰 GUI。

[6] When 引导的是时间状语从句，主句中主语为 an additional list… or a dialog box，谓语为 may appear；that 引导的是定语从句。

[7] that 引导的是定语从句，从句中 it 是形式宾语，真正的宾语是 to find and use…。

[8] ribbon，在基于 GUI 的应用软件中，ribbon 是一种接口。在这种接口中，一组工具栏放在一个标记条（tab bar）中。在微软的某些应用软件中，这种格式纳入模块化 ribbon，作为主接口。contextual tabs，在微软 ribbon framework 应用中，它是一个隐藏的标记控件，当在应用工作区中，一个对象，如一幅图像被选中或呈高亮状态时，该控件就会显示在标记行中。

[9] alternatives 为多种选择，此处是指很多图形。

KEYWORDS

template	模（样）板
option	选项
menu	菜单
menu bar	菜单栏
pointer	指针
window	窗口
dialog box	对话框
toolbar	工具栏

button	按钮
list of menu	菜单列表
shortcut	快捷（方式）
ribbon	条形框
tab	选项卡
contextual tab	上下文选项卡
gallery	图库
presentation	显示，表示，呈现，显现

EXERCISES

Multiple Choices

1. Basic concepts and functions of the office automation software include_____.
   a. common document-handling tasks   b. software suite
   c. ownership rights   d. how to get help
2. The following concepts for document-handling are universal_____.
   a. opening a document   b. saving a document
   c. turning on a computer   d. printing a document
3. Microsoft Office includes the following software_____.
   a. Word   b. Excel   c. PowerPoint   d. Access
4. If we want to insert a spreadsheet table into a letter written by using the Word, we should _____.
   a. not be able to use the Excel   b. be able to use the Excel
   c. locate the particular table   d. copy and paste the table into our letter
5. Online help is a tool that_____.
   a. provides online assistance
   b. provides help with forcing you to leave your computer screen
   c. provides help without forcing you to leave your computer screen
   d. has a table-of-contents feature
6. Proprietary software can be_____.
   a. a system software   b. an application program
   c. owned by its maker only   d. owned by its buyer
7. Although shareware is a software that you can install and use, but most shareware specifies that_____.
   a. you need to pay for a trial period   b. you don't need to pay for a trial period
   c. you need to pay for a long term   d. you don't need to pay for a long term
8. Freeware is a_____.
   a. public-domain software
   b. program you can share with others free of charge

c. program you can use after buying it

d. program you can use free of charge

9. Icons _____.

   a. are graphical elements

   b. are used to represent familiar objects and a mouse

   c. can be selected by a pointer which is controlled by a mouse

   d. can be used to replace commands

10. A window is _____.

    a. simply a rectangular area

    b. one type of Windows operating system

    c. used to contain a document, program, or massage

    d. used to display some objects or elements

11. Microsoft Office 2010 has _____.

    a. menus   b. dialog boxes   c. toolbar   d. buttons

12. The newest Office version introduces _____.

    a. ribbons   b. contextual tabs   c. galleries   d. others

8.2 MICROSOFT OFFICE 2013

Microsoft Office 2013 (formerly Office 15) is a version of Microsoft Office, a productivity suite for Microsoft Windows. It is the successor of Microsoft Office 2010 and includes extended file format support, user interface updates and support for touch among its new features. Office 2013 is suitable for IA-32 and x64 systems and requires Windows 7, Windows Server 2008 R2 or a later version of either [1]. A version of Office 2013 comes included on Windows RT devices [2].

Microsoft Office 2013 comes in twelve different editions, including three editions for retail outlets, two editions for volume licensing channel, five subscription-based editions available through Microsoft Office 365 program, the web application edition known as Office Web Apps and the Office RT edition made for tablets and mobile devices [3].

On February 25, 2014, Microsoft Office 2013 Service Pack 1 (SP1) was released [4].

1. New features

Office 2013 is more cloud-based than previous versions; a domain login, Office 365 account, or Microsoft account can now be used to sync Office application settings (including recent documents) between devices, and users can also save documents directly to their SkyDrive account [5].

New features include a new read mode in Microsoft Word, a presentation mode in Microsoft PowerPoint and improved touch and inking in all of the Office programs. Microsoft

Word can also insert video and audio from online sources as well as the capability to broadcast documents on the Web. Word and PowerPoint also have bookmark-like features which sync the position of the document between different computers.

The Office Web Apps suite was also updated for Office 2013, introducing additional editing features and interface changes.

Other features of Office 2013 include:

- Flatter look of the Ribbon interface and subtle animations when typing or selecting (Word and Excel)
- A new visualization for scheduled tasks in Microsoft Outlook
- Remodeled start screen
- New graphical options in Word
- Objects such as images can be freely moved; they snap to boundaries such as paragraph edges, document margin and or column boundaries
- Online picture support with content from Office.com, Bing.com and Flickr (by default, only images in public domain)[6]
- Ability to return to the last viewed or edited location in Word and PowerPoint
- New slide designs, animations and transitions in PowerPoint 2013
- Support for Outlook.com and Hotmail.com in Outlook
- Support for integration with Skype, Yammer and SkyDrive[7]
- IMAP special folders support[8]

Figure 8-5 shows the lineup of Microsoft Office 2013 icons.

Figure 8-5 Lineup of Microsoft Office 2013 icons, from left to right: Word, Excel, PowerPoint, Outlook, Access, OneNote, Publisher, Lync and InfoPath

2. Editions

As with previous versions, Office 2013 is made available in several distinct editions aimed towards different markets. All traditional editions of Microsoft Office 2013 contain Word, Excel, PowerPoint and OneNote and are licensed for use on one computer.

Five traditional editions of Office 2013 were released:

- Home & Student: This retail suite includes the core applications Word, Excel, PowerPoint, and OneNote[9].
- Home & Business: This retail suite includes the core applications Word, Excel, PowerPoint, and OneNote plus Outlook[10].
- Standard: This suite, only available through volume licensing channels, includes the core applications Word, Excel, PowerPoint, and OneNote plus Outlook and

Publisher[11].
- Professional: This retail suite includes the core applications Word, Excel, PowerPoint, and OneNote plus Outlook, Publisher and Access[12].
- Professional Plus: This suite, only available through volume licensing channels, includes the core applications Word, Excel, PowerPoint, and OneNote plus Outlook, Publisher, Access, InfoPath and Lync[13].

NOTES

[1] IA-32（Intel Architecture）即英特尔体系架构，属于 x86 体系结构的 32 位版本，即具有 32 位内存地址和 32 位数据操作数的处理器体系结构，从 1985 年面世的 80386 直到 Pentium 4，都是使用 IA-32 体系结构的处理器。x64 是指 64 位操作系统，在计算机架构中，64 位整数、内存地址或其他数据单元，是指它们最高达到 64 位（8 字节）宽。Windows Server 2008 R2 是一款服务器操作系统。和 2008 年 1 月发布的 Windows Server 2008 相比，Windows Server 2008 R2 继续提升了虚拟化、系统管理弹性、网络存取方式，以及信息安全等领域的应用，其中有不少功能需要搭配 Windows 7 操作系统。

[2] included on…，过去分词短语作状语。Windows RT 是 Windows 家族的一个新成员，新系统画面与操作方式变化极大，采用全新的 Metro（新 Windows UI）风格用户界面，各种应用程序、快捷方式等能以动态方块的样式呈现在屏幕上，用户可自行将常用的浏览器、社交网络、游戏、操作界面融入。

[3] 长句，介绍 Office2013 的 12 种版本。Office 365 是微软带给所有企业最佳生产力和高效协同的高端云服务，是微软公司基于云平台的应用套件。Office Web Apps 是由微软推出的基于 Web 端的在线办公工具，它将 Microsoft Office 2010 产品的体验延伸到可支持的浏览器上。Office RT 即 Office 2013 RT 版，是微软为 ARM 设备量身打造的，所有搭载 Windows RT 操作系统的平板等都预装 Office 2013 RT，包括 Surface 平板电脑。

[4] Service Pack1, Service Pack 直译是服务包,操作系统中比较大的而且重要的升级补丁，一般说法是补丁，用途是修补系统、大型软件中的安全漏洞，一般是补丁的集合，简称 SP。

[5] SkyDrive 是微软公司推出的一项云存储服务,用户可以通过自己的 Windows Live 账户进行登录，上传自己的图片、文档等到 SkyDrive 中进行存储。

[6] Bing,微软必应搜索是国际领先的搜索引擎，为中国用户提供网页、图片、视频、词典、翻译、资讯、地图等全球信息搜索服务。Flickr 的注释详见 5.1.2 节的 NOTES[5]。

[7] Skype 是一款即时通信软件，它具备 IM 所需的功能，比如视频聊天、多人语音会议、多人聊天、传送文件、文字聊天等功能。它可以免费高清晰与其他用户语音对话，也可以拨打国内国际电话，无论固定电话、手机、小灵通均可直接拨打，并且可以实现呼叫转移、短信发送等功能。

[8] IMAP（Internet Mail Access Protocol）即 Internet 邮件访问协议，以前称作交互邮件访问协议（Interactive Mail Access Protocol），它的主要作用是邮件客户端（例如 MS Outlook Express）可以通过这种协议从邮件服务器上获取邮件信息，下载邮件等。

[9] OneNote，是一种数字笔记本，它为用户提供了一个收集笔记和信息的位置，并提供了强大的搜索功能和易用的共享笔记本。

[10] Outlook 是微软办公软件套装的组件之一，它对 Windows 自带的 Outlook express 的功能进行了扩充。Outlook 的功能很多，可以用它来收发电子邮件、管理联系人信息、记日记、安排日程、分配任务。目前最新版为 Outlook 2013。

[11] Publisher 是 Microsoft Office 组件之一。 Publisher 是完整的企业发布和营销材料的解决方案。

[12] Access 是由微软发布的关系数据库管理系统。它结合了 Microsoft Jet Database Engine 和图形用户界面两项特点，是 Microsoft Office 的系统程序之一。

[13] InfoPath 是企业级搜集信息和制作表单的工具，将很多的界面控件集成在该工具中，为企业开发表单搜集系统提供了极大的方便。Lync 是微软带有即时消息、会议和语音功能的真正的统一通信客户端软件。

KEYWORDS

productivity	生产力，生产率
suite	一组，一套，套件，程序序列
touch	触摸，接触
retail outlet	零售商店
licensing	许可，特许
subscription	预定，预约，预约费
cloud-based	基于云的
presentation	显示，表示，呈现，显现
domain	域
login	注册，挂号，登录，进入系统
setting	设置，调整，装置
inking	着墨，给……上油墨
ribbon interface	带状界面
icon	图标，图符，图示
typing	打印，打字
visualization	可视化，目视，显像
remodel	改做，改造，改……之型
snap	抓取，快照，按钮接头，快速移动
animation	动画制作，直观显示
transition	转换，转移，变迁，过渡
folder	文件夹，信息页面

CHAPTER 8 OFFICE AUTOMATION SOFTWARE

EXERCISES

Fill in the blanks with appropriate words or phrases found behind this exercises.

1. Microsoft Office 2013 is the _____ of Microsoft office 2010.
2. Microsoft Office 2013 has the following new features: _____ .
3. Microsoft Office 2013 requires using _____ .
4. Microsoft Office 2013 has _____ editions.
5. IA (Intel Architecture)-32 belongs to 32-bit edition of _____ .
6. Microsoft Office 2013 has three editions for _____ .
7. If we want to use the subscription-based editions of Microsoft Office 2013, we should have aid of _____ .
8. Office Web Apps is the _____ edition.
9. Office 365 is an application suite of _____ .
10. In Office 2013 users can save documents directly to their _____ account.
11. We can insert video and audio from _____ into Microsoft Word.
12. Microsoft Outlook has a new feature of _____ for scheduled tasks.
13. IMAP is an abbreviation for _____ .
14. Comparing edition of Home & Student with edition of Home & Business, we found that the successor has one more feature, it is _____ .
15. Through volume licensing channels we can use _____ editions.
16. The last edition of outlook is _____ .

   a. x86 architecture
   b. Outlook
   c. SkyDrive
   d. successor
   e. web application
   f. Outlook 2013
   g. Windows 7
   h. visualization
   i. Microsoft Office 365 program
   j. Extended file format, user interface updates and touch support
   k. Standard and Professional Plus
   l. on line sources
   m. retail outlets
   n. cloud-based platform
   o. 12
   p. Internet Mail Access Protocol

CHAPTER 9　MULTIMEDIA

9.1　MULTIMEDIA AND ITS MAJOR CHARACTERISTICS

1. Definition

Multimedia refers to content that uses a combination of different content forms. This contrasts with media that use only rudimentary computer displays such as text-only or traditional forms of printed or hand-produced material. Multimedia includes a combination of text, audio, still images, animation, video, or interactive content forms, as shown in Figure 9-1.

Figure 9-1　Examples of individual content forms combined in multimedia[1]

Multimedia can be recorded and played, displayed, dynamic, interacted with or accessed by information content processing devices, such as computerized and electronic devices, but can also be part of a live performance [2]. Multimedia devices are electronic media devices used to store and experience multimedia content. Multimedia is distinguished from mixed media in fine art; by including audio, for example, it has a broader scope. The term "rich media" is synonymous with interactive multimedia. Hypermedia scales up the amount of media content in multimedia application.

2. Categorization of multimedia

Multimedia may be broadly divided into linear and non-linear categories. Linear active content progresses often without any navigational control for the viewer such as a cinema presentation[3]. Non-linear uses interactivity to control progress as with a video game or self-paced computer based training. Hypermedia is an example of non-linear content.

3. Major characteristics of multimedia

Multimedia presentations may be viewed by person on stage, projected, transmitted, or played locally with a media player. A broadcast may be a live or recorded multimedia presentation. Broadcasts and recordings can be either analog or digital electronic media technology. Digital online multimedia may be downloaded or streamed. Streaming multimedia may be live or on-demand.

Multimedia games and simulations may be used in a physical environment with special effects, with multiple users in an online network, or locally with an offline computer, game system, or simulator [4].

The various formats of technological or digital multimedia may be intended to enhance the users' experience, for example to make it easier and faster to convey information. Or in entertainment or art, to transcend everyday experience [5].

Enhanced levels of interactivity are made possible by combining multiple forms of media content. Online multimedia is increasingly becoming object-oriented and data-driven, enabling applications with collaborative end-user innovation and personalization on multiple forms of content over time [6]. Examples of these ranging from multiple forms of content on Web sites like photo galleries with both images (pictures) and title (text) user-updated, to simulations whose events, illustrations, animations or videos are modifiable, allowing the multimedia "experience" to be altered without reprogramming [7]. In addition to seeing and hearing, haptic technology enables virtual objects to be felt. Emerging technology involving illusions of taste and smell may also enhance the multimedia experience.

NOTES

[1] video footage，视频片段，视频画面，视频录像，录像（见图9-1）。

[2] 长句，句型为 Multimedia can be…but can also be…，中间插入由 such as 引导的同位语。

[3] progress 此处为动词。

[4] 长句。句中有3个 with 引导的方式状语。

[5] 由于紧跟前一句，此句省略了主语和谓语。

[6] 长句。enabling…，现在分词短语作状语。

[7] 长句。主句结构为 Examples of…are modifiable，其中 ranging from…, to…为分词短语作定语，而 allowing…为现在分词短语作状语。

KEYWORDS

rudimentary	基本的，初步的
animation	动画
rich media	富媒体，多元媒体

synonymous	同义的，同义词的
hypermedia	超媒体
convey	传达，运输
collaborative	合作的，协作的
simulation	模拟，仿真
streaming multimedia	流式媒体，流媒体
live	实况，直播
object-oriented	面向对象的
data-driven	数据驱动的
haptic	触觉的

EXERCISES
Multiple choices
1. Multimedia refers to use _____.
    a. only rudimentary computer displays   b. a combination of different content forms
    c. a single content form only   d. many kinds of digital medium
2. Multimedia includes a combination of _____.
    a. text   b. still images   c. video   d. interactive content forms
3. Multimedia devices are _____ devices.
    a. computerized b. electronic   c. analog only   d. digital
4. There are several multimedia forms in this text, they are _____ media.
    a. super   b. hyper   c. rich   d. mixed
5. Multimedia can be _____ by information content processing devices.
    a. recorded   b. played   c. displayed   d. accessed
6. The content forms in multimedia can be _____.
    a. mouse   b. video footage   c. animation   d. audio
7. Non-linear multimedia _____.
    a. uses interactivity to control progress
    b. does not need any navigational control when its active content progresses
    c. has an example of hypermedia
    d. has a typical application which is video game
8. Multimedia presentations may be _____.
    a. transmitted   b. projected   c. played   d. viewed
9. Right now emerging technology in multimedia involves _____.
    a. seeing   b. hearing   c. tasting   d. smelling
10. Online multimedia will enable _____.
    a. applications with collaborative end-user innovation
    b. applications with personalization

c. the multimedia "experience" to be altered without reprogramming

d. virtual objects to be felt

9.2 USAGE/APPLICATION

Multimedia finds its application in various areas including, but not limited to, advertisements, art, education, entertainment, engineering, medicine, mathematics, business, scientific research and spatial temporal applications. Figure 9-2 shows a presentation using PowerPoint.Figure 9-3 shows a virtual reality using multimedia content. Figure 9-4 shows a multimedia-terminal in Dresden (Germany).

Figure 9-2   A presentation using PowerPoint. Corporate presentations may combine all forms of media content

Figure 9-3   Virtual reality uses multimedia content. Applications and delivery platforms of multimedia are virtually limitless

Several examples are as follows:

1. Creative industries

Creative industries use multimedia for a variety of purposes ranging from fine arts, to entertainment, to commercial art, to journalism, to media and software services provided for any of the industries listed below[1]. An individual multimedia designer may cover the spectrum throughout their career. Requests for their skills range from technical, to analytical, to creative.

Figure 9-4   Multimedia-Terminal in Dresden (Germany)

2. Commercial uses

Much of the electronic old and new media used by commercial artists and graphic designers is multimedia. Exciting presentations are used to grab and keep attention in advertising. Business to business, and interoffice communications are often developed by creative services firms for advanced multimedia presentations beyond simple slide shows to sell ideas or liven-up training[2]. Commercial multimedia developers may be hired to design for governmental services and nonprofit services applications as well.

3. Entertainment and fine arts

In addition, multimedia is heavily used in the entertainment industry, especially to develop special effects in movies and animations (VFX, 3D animation, etc.)[3]. Multimedia games are a popular pastime and are software programs available either as CD-ROMs or online. Some video games also use multimedia features. Multimedia applications that allow users to actively participate instead of just sitting by as passive recipients of information are called Interactive Multimedia[4]. In the Arts, there are multimedia artists, whose minds are able to blend techniques using different media that in some way incorporates interaction with the viewer[5].

4. Education

In Education, multimedia is used to produce computer-based training courses (popularly called CBTs) and reference books like encyclopedia and almanacs[6]. A CBT lets the user go through a series of presentations, text about a particular topic, and associated illustrations in various information formats. Edutainment is the combination of education with entertainment, especially multimedia entertainment.

Learning theory in the past decade has expanded dramatically because of the introduction of multimedia. Several lines of research have evolved (e.g. Cognitive load, Multimedia learning, and the list goes on)[7]. The possibilities for learning and instruction are nearly endless.

5. Journalism

Newspaper companies all over are also trying to embrace the new phenomenon by implementing its practices in their work [8]. While some have been slow to come around, other major newspapers like The New York Times, USA Today and The Washington Post are setting the precedent for the positioning of the newspaper industry in a globalized world[9].

News reporting is not limited to traditional media outlets. Freelance journalists can make use of different new media to produce multimedia pieces for their news stories. It engages global audiences and tells stories with technology, which develops new communication techniques for both media producers and consumers.

6. Engineering

Software engineers may use multimedia in Computer Simulations for anything from entertainment to training such as military or industrial training. Multimedia for software interfaces are often done as a collaboration between creative professionals and software engineers.

7. Industry

In the Industrial sector, multimedia is used as a way to help present information to shareholders, superiors and coworkers. Multimedia is also helpful for providing employee training, advertising and selling products all over the world via virtually unlimited web-based technology.

8. Mathematical and scientific research

In mathematical and scientific research, multimedia is mainly used for modeling and simulation. For example, a scientist can look at a molecular model of a particular substance and manipulate it to arrive at a new substance.

9. Medicine

In Medicine, doctors can get trained by looking at a virtual surgery or they can simulate how the human body is affected by diseases spread by viruses and bacteria and then develop techniques to prevent it[10]. Multimedia application like virtual surgeries also help doctors to get practical training.

NOTES

[1] 长句，ranging from…, to…, to…，现在分词短语作定语。fine arts，美艺术（绘画、雕塑、音乐、舞蹈、文学、工艺、建筑等）。

[2] Business to business 即 B2B，电子商务中的商业对商业的运作模式，详见 4.3.1 节。beyond…，介词短语作状语。 liven-up，使有生气，活跃气氛。

[3] VFX（Visual Effects），视觉效果。3D animation，三维动画。

[4] 长句，句子结构为 Multimedia applications…are called…，中间有 that 引导的定语从句以及 instead of…，介词短语，意为"而不是……"。

[5] 长句,主句为 whose minds…,句中 using…现在分词短语作定语,修饰 techniques,而后面的 that 引导的定语从句，修饰 media。

[6] CBT（Computer-Based Training）即计算机辅助训练，广泛应用于培训领域，如驾驶、医疗等。CBT 培训是采用图像、声音、文字、仿真模拟操作等多媒体手段为学员营造直观、真实的培训环境，并提供接近真实的互动操作，使学员在培训过程中有身临其境的感受，增强培训效果。

[7] cognitive load，认知负荷，来源于认知负荷理论（cognitive load theory），是 Sweller 等人在 20 世纪 80 年代提出的，主要从认知资源分配的角度考察学习和问题的解决。the list goes on，意为"还有很多"。

[8] phenomenon，此处是指多媒体这一特殊事务。

[9] 长句。While 引导的是并列分句，主句是 other major newspapers…are setting…。其中 setting the precedent for…为创立……的先例（先机）。

[10] 由 or 连接的并列句,后一句中又是一个由 and 连接的并列句,句子结构为 they can simulate… and then develop…，其中 how 引导的是宾语从句。

KEYWORDS

advertisement	广告，宣传，启事
entertainment	娱乐，招待，表演
spatial	空间的
temporal	时间的
journalism	新闻学（界），报刊编辑
spectrum	光谱，范围

presentation	显示，表示，呈现，显现
slide	幻灯片
animation	动画
illustration	插画，插图，实例
edutainment	教育娱乐，寓教于乐
phenomenon	现象，特殊的事务
pastime	娱乐，消遣
almanacs	年鉴
cognitive	认知的
modeling	建造模型（建模），造型
simulation	模拟

EXERCISES

Fill in the blanks with appropriate words or phrases found behind this exercise.

1. An individual multimedia _____ may cover the spectrum throughout their career.
2. Exciting presentations are used to grab and keep attention in_____.
3. Multimedia games are a popular pastime and are software programs available either as CD-ROMs or_____.
4. Learning theory in the past decade has expanded dramatically because of the introduction of_____.
5. Freelance_____ can make use of different new media to produce multimedia pieces for their news stories.
6. In mathematical and scientific research, multimedia is mainly used for _____ and simulation.
7. Multimedia application like virtual surgeries also help _____ to get practical training.
8. Software engineers may use multimedia in _____.
9. Multimedia in the industries is helpful for providing employee _____.
10. Corporate presentations may combine all forms of _____.
11. Applications and delivery platforms of multimedia are _____.
12. In this text we present _____ kinds of multimedia applications.

    a. journalists
    b. doctors
    c. advertising
    d. computer simulations
    e. multimedia
    f. designer
    g. training, advertising and selling products

h. nine
i. online
j. virtually limitless
k. media content
l. modeling

CHAPTER 10
COMPUTER GRAPHICS AND IMAGES

10.1 THE VARIOUS COMPUTER GRAPHICS

1. Two-dimensional computer graphics

2D computer graphics are the computer-based generation of digital images—mostly from models, such as digital image, and by techniques specific to them.

2D computer graphics are mainly used in applications that were originally developed upon traditional printing and drawing technologies such as typography. In those applications, the two-dimensional image is not just a representation of a real-world object, but an independent artifact with added semantic value; two-dimensional models are therefore preferred, because they give more direct control of the image than 3D computer graphics, whose approach is more akin to photography than to typography [1].

(1) Pixel art

A large form of digital art being pixel art is created through the use of raster graphics software, where images are edited on the pixel level[2]. Graphics in most old (or relatively limited) computer and video games, graphing calculator games, and many mobile phone games are mostly pixel art.

(2) Sprite graphics

A sprite is a two-dimensional image or animation that is integrated into a larger scene. Initially including just graphical objects handled separately from the memory bitmap of a video display, this now includes various manners of graphical overlays [3].

Originally, sprites were a method of integrating unrelated bitmaps so that they appeared to be part of the normal bitmap on a screen, such as creating an animated character that can be moved on a screen without altering the data defining the overall screen[4]. Such sprites can be created by either electronic circuitry or software. In circuitry, a hardware sprite is a hardware construct that employs custom DMA channels to integrate visual elements with the main screen in that it super-imposes two discrete video sources[5]. Software can simulate this through specialized rendering methods.

(3) Vector graphics

Vector graphics formats are complementary to raster graphics. Raster graphics is the representation of images as an array of pixels and is typically used for the representation of photographic images. Vector graphics consists of encoding information about shapes and colors that comprise the image, which can allow for more flexibility in rendering. There are instances when working with vector tools and formats is best practice, and instances when working with raster tools and formats is best practice[6]. There are times when both formats come together. An understanding of the advantages and limitations of each technology and the relationship between them is most likely to result in efficient and effective use of tools.

Figure 10-1 shows the effect of vector graphics versus raster (bitmap) graphics.

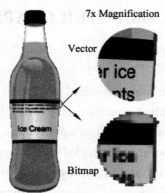

Figure 10-1　Example showing effect of vector graphics versus raster (bitmap) graphics

2. Three-dimensional computer graphics

3D graphics compared to 2D graphics are graphics that use a three-dimensional representation of geometric data. For the purpose of performance, this is stored in the computer[7]. This includes images that may be for later display or for real-time viewing.

Despite these differences, 3D computer graphics rely on similar algorithms as 2D computer graphics do in the frame and raster graphics (like in 2D) in the final rendered display[8]. In computer graphics software, the distinction between 2D and 3D is occasionally blurred; 2D applications may use 3D techniques to achieve effects such as lighting, and primarily 3D may use 2D rendering techniques.

3D computer graphics are the same as 3D models. The model is contained within the graphical data file, apart from the rendering. However, there are differences that include the 3D model is the representation of any 3D object. Until visually displayed a model is not graphic. Due to printing, 3D models are not only confined to virtual space. 3D rendering is how a model can be displayed. Also can be used in non-graphical computer simulations and calculations[9].

3. Computer animation

Computer animation is the art of creating moving images via the use of computers. It is a subfield of computer graphics and animation. Increasingly it is created by means of 3D computer graphics, though 2D computer graphics are still widely used for low bandwidth, and faster real-time rendering needs[10]. Sometimes the target of the animation is the computer itself, but sometimes the target is another medium, such as film. It is also referred to as CGI (Computer-generated imagery or computer-generated imaging), especially when used in films.

Figure 10-2 shows an example of computer animation produced using Motion capture.

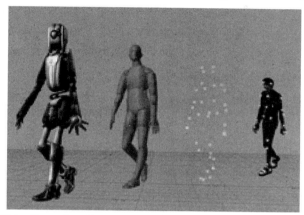

Figure 10-2　Example of Computer animation produced using Motion capture

NOTES

[1] 分号隔开的两个句子，后一句中 because…是原因状语从句，whose approach…为非限定性定语从句。

[2] being pixel art，现在分词短语作定语，修饰 digital art，where 引导的是非限定性定语从句。

[3] 长句。其中 including…现在分词短语作状语。sprite 在国内很多人叫精灵，是一种网页图片应用处理方式。它允许将一个页面涉及的所有零星图片都包含到一张大图中去，这样一来，当访问该页面时，载入的图片就不会像以前那样一幅一幅地慢慢显示出来了。

[4] 长句。so that…目的状语从句，such as…为同位语，其中 that 引导的定语从句，修饰 character（角色）。without altering…现在分词短语作状语，而 defining…为现在分词短语作定语，修饰 data。

[5] 长句。第一个 that 引导的定语从句，修饰 construct，第二个 in that…为方式状语从句。

[6] 句子结构为 There are instances… and instances…。

[7] performance，性能，此处意为展现，呈现。this 为 3D 图形数据。

[8] as 2D computer graphics do…，方式状语从句。

[9] Also 后面省略了主语 3D rendering。

[10] 长句，though 引导的是让步状语从句。

KEYWORDS

photography	摄影，照相
typography	排版
pixel	像素
artifact	人工制品
semantic	语义的
raster graphics	光栅图形（像）
sprite	精灵
bitmap	位图
vector	矢量，向量
encoding	编码
DMA (Direct Memory Access)	直接存储器访问
animation	动画
algorithm	算法
blur	模糊
stylistic	文体的，风格上的，格式上的
frame	帧

EXERCISES

Fill in the blanks with appropriate words or phrases found behind this exercises.

1. 2D computer graphics are mainly used in applications that were originally developed upon traditional printing and drawing technologies such as_____.
2. A large form of digital art being pixel art is created through the use of _____software, where images are edited on the pixel level.
3. A sprite is a two-dimensional image or animation that is_____ into a larger scene.
4. Vector graphics consists of encoding information about shapes and colors that comprise the image, which can allow for more flexibility in_____.
5. 3D rendering is how a _____can be displayed.
6. Vector graphics formats are_____ to raster graphics.
7. Two-dimensional models give more direct control of the image than _____.

   a. model
   b. integrated
   c. typograph
   d. raster graphics
   e. complementary
   f. rendering
   g. 3D computer graphics

10.2　GRAPHICS SOFTWARE (1)

1. Desktop Publishing

Desktop publishing (DTP) grew naturally out of word processing though for a long time it was a separate activity[1]. Recently the two have shown all the signs of growing back together again. The difference between basic word processing (WP) and DTP can be seen by considering the traditional function of the author of a printed document, as compared with the function of the printer[2]. Before the advent of the desktop computer the author was responsible for producing a typescript—the process of assembling words in the right order. The printer then took those words and (perhaps with the aid of a designer or typographer) laid out the text in a particular manner, with or without appropriate illustrations, and printed them. The modern author does exactly what his predecessor did, but using a word processor, so that the words do not have to be retyped by the printer[3]. What DTP does is to automate most of the functions of the printer, using a desktop computer[4]. Four developments made this possible: the desktop computer with a GUI, DTP software, the laser printer and the page-description language-the PDL.

The importance of the laser printer was that a high-quality final product could be produced without the need for traditional typesetting processes (particularly the use of moveable type). The early laser printers, operating at 300~400dots per inch (dpi) could not rival traditional printing processes but could produce results that were acceptable for many everyday purposes, and at a much lower price[5]. The development of higher resolution laser printers and the digital type-setter (now called an image-setter) have since made it possible for work of almost any quality to be produced this way, though the term DTP, in some people's minds, is still associated with a poorer quality, amateur product [6].

The importance of the GUI lays in the fact that the function of DTP software is to lay out pre-prepared text and graphics and a GUI allows the user to see immediately an accurate representation of the final product[7]. This "what you see is what you get" (WYSIWYG) feature is vital to DTP though the slogan should be taken with a pinch of salt[8]. "what you see" on a 72 dpi screen can never be a wholly accurate representation of "what you get" on a 300 ~ 1200dpi (or better) printer. The slogan should, perhaps, be changed to WYSIANATTCMTWYG—"what you see is as near as the technology can manage to what you get". A necessary feature of a DTP package, therefore, is a zoom facility which displays a portion of the document at a larger size than normal so as to display it at something much closer to the resolution of the final printed product[9]. Unfortunately only a small part of the document can be seen at any one time in this magnified mode.

2. Electronic Publishing (CD-ROMs and the Internet)

In recent years more and more material has been published electronically rather than on paper. The two most important new media have been CD ROM and the Internet. As a result, modern versions of many DTP packages and word processors provide the facility to output files in the formats which have been developed specifically for these two media—particularly HTML (Hyper-Text Markup Language) and PDF (Portable Document Format).

The hyper-text principle has also been extended to links to photographs, drawings, sounds, video, animations, tables of figures, maps etc, and the benefit to the user is that vast amounts of information can be made available—in such a way that the user can decide how much to access and in what order[10]. Print publishing on the other hand is still, essentially, a linear process. The reader reads the material in the order dictated by the writer, starting at the beginning and going on to the end. Hyperlinked material has no unique beginning and no unique end.

All this means that electronic publishing, though having many obvious overlaps with paper publishing, is very different medium needing different skills and different software[11].

New types of graphic-oriented software have been designed to produce electronic publications: web-publishing software; web-graphic design software; multimedia "authoring" software; PDF publishing software. One of the potential problems in establishing new software products of this kind is that of the proliferation of standards. Two standards, already referred to, are the PDF format—established by Adobe through its "Acrobat" suite of software. The other is the HTML format, now in the process of being extended to XML (Extensible ML), DHTML (Dynamic HTML) and VRML (Virtual Reality ML).

NOTES

[1] 主从复合句，Desktop…processing 是主句，though 引导一个让步状语从句，for a long time 是从句中的时间状语。

[2] The difference… and DTP 是主语，as…引导的是分词短语作状语。

[3] The modern…did 是主句，其中 what…是宾语从句，so that 引导一个目的状语从句。

[4] What 引导的是主语从句。

[5] The early…results 是主句，that…为定语从句，主句中 operating…是分词短语作定语。

[6] The development of…this way 为主句，though…product 为让步状语从句，in some people's minds 是从句中的插入语；since 为副词，意为从那时起；it 为先行代词，代替 to be produced…作宾语。

[7] that the function…and a GUI…product 是并列的同位语从句。

[8] with a pinch of salt 意为有些保留，though 引导的是让步状语从句。

[9] which 引导一个定语从句，so as to 引导一个目的状语。

[10] 由 and 连接的两个并列句，后一句中 that 引导的是表语从句；破折号后面是同位语，其中 that 引导的是状语从句。

[11] 长句，that 引导的是宾语从句，中间插入 though 引导的让步状语从句，needing…分词短语作定语，overlap 重叠，重复。

KEYWORDS

GUI (Graphics User Interface)	图形用户界面
DTP (Desktop Publishing)	桌面出版
resolution	分辨率
image-setter	激光照排机
paste-board	粘贴板
HTML (Hypertext Markup Language)	超文本标记语言
PDF (Portable Document Format)	可移植文档格式
PDL (Page Description Language)	页面描述语言
graphics software	图形软件
WP(Word Processing)	（文）字处理
typescript	打印文稿
word processor	字处理器（软件）
laser printer	激光打印机
dpi (dots per inch)	每英寸点数
Electronic Publishing	电子出版
video	视频
animation	动画
hyperlink	超链接
suite of software	软件套件

EXERCISES

1. Single choice (according to the text).

(1) DTP is_____.
    a. Digital transmitting processor   b. Desktop publishing
    c. Data transferring pipe

(2) GUI is_____.
    a. General User Interface   b. Graphics User Input
    c. Graphics User Interface

(3) WYSIWYG means_____.
    a. What you see is what you get   b. Where you see is where you get
    c. When you see is when you get

(4) PDF is_____.
   a. Portable Document Format   b. Programming Data Format
   c. Prolog Document Format
(5) HTML is_____.
   a. Hypertext Markup Logic   b. Hypertext Markup Language
   c. Hypertext Markup Link
(6) PDL is_____.
   a. Push Down List   b. Page Description Language
   c. Programmable Digital Logic
(7) Recently _____ have shown all the signs of growing back together again.
   a. WP and GUI   b. WP and CD-ROM
   c. WP and DTP
(8) _____ is still a linear process.
   a. Hyperlinked material   b. Print publishing
   c. Web-oriented reading

2. True/False.

(1) _____ The importance of the laser printer was that a high-quality final product could be produced without the need for traditional typesetting processes.
(2) _____ The two most important new media are CD-ROM and Internet.
(3) _____ Acrobat is the product of Microsoft.
(4) _____ HTML can also be thought of as a PDL which enables the user to specify what a Web page should look like and how it should link to other pages.
(5) _____ PDF format was designed to describe the appearance of a Web page.
(6) _____ Electronic publishing has many obvious overlaps with paper publishing.
(7) _____ PDF is an extension of Adobe's PDL.
(8) _____ The early laser printers can operate at 300-1200 dpi.

10.3 GRAPHICS SOFTWARE (2)

1. Computer-aided design (CAD)

CAD is a classic example of a graphic application which has grown up on large powerful computers and at one time it would have been considered quite impractical to do any serious work of this type on a desktop computer[1]. It is true that much important CAD work is still done on graphic workstations but increasingly powerful applications are now also being run on ordinary desktop computers. One of the simplest types of computer-aided design is the preparation of 2-D plans—the sort of work done traditionally by draughtsman on a drawing board with the aid of tee-squares and compasses. This type of work can be done very rapidly with software which provides an electronic tee-square, rubber-banding, standardized parts and geometrical shapes, "instancing", grids, layers and a library of standard shapes[2].

The more powerful CAD packages can be used to design and represent 3-D objects. These are sometimes called solid-modelers because they create a complete and unambiguous description of a 3-D object in the computer and can display images with hidden lines removed and exterior surfaces shaded. Because of the wide-spread interest in 3-D modeling, which has now spread well beyond the boundaries of CAD.

2. Illustration (drawing) software

Though CAD software is associated with engineering design and other forms of technical drawing the facilities offered by such packages have proved useful in a wide variety of other applications. Illustration (or simply Drawing) software can be thought of as CAD software with a less specialized purpose. Like CAD software the shapes that are drawn are created and stored as vectors and so this type of software should be distinguished from "Painting software" which is used to create and modify bit-mapped images[3]. The differences between CAD and Illustration software lie in emphasis and direction rather than principle. In both types of software the user can assemble and edit a range of line-based geometric shapes; objects can be separately selected, moved, distorted and re-sized; bounded shapes can be filled with colours and patterns; objects can be grouped and treated as single objects and a variety of positioning aids like grids and snapping are available. The differences lie in a number of enhancements required, and a number of specialist features not required, by the typical user who might be using the software to produce semi-technical illustrations but might equally be producing nontechnical drawings for advertisements, leaflets and magazines[4]. In general such applications require a larger range of fonts, more colour and colour-effects, a completely different library of ready-made drawings and a more sophisticated range of curves. This images that are to be produced by this type of drawing package are less severely functional and more aesthetically pleasing—more like "art".

3. Business presentation software

This is another type of vector software with a very focused application—the rapid production of "presentations" (informative talks or short articles) which summarize as compactly as possible a set of ideas and/or data. The main features are simple drawing facilities (like illustration software), clip-art, attractively presented, small quantities of text and graphs. The target user, unlike the user of Illustration software or CAD, is not a designer and therefore it is important to be able to automate the production of an aesthetically-pleasing result. This is done by the provision of a wide range of "templates". A template, in the ordinary use of the term, is a pattern or guide that is used to help in the creation of some object—like a plastic sheet with the shape of letters of the alphabet cut in it which can be used to rapidly trace letters onto paper[5]. In graphics software a template is a design which the user can quickly adapt to their own purposes. For example, it could be a screen of "dummy" text which users adapt by substituting their own text for the one provided[6]. The point of the exercise is that a professional-graphic designer is employed in creating the templates—in choosing suitable fonts and colors, perhaps with a patterned background—and so the user does not need any design skills to produce a professional-looking result[7]. The process is also a lot quicker for the users than creating the screen from scratch.

NOTES

[1] 长句，which 引导的是定语从句；it 为先行代词，逻辑主语为 to do…；at one time 意思是曾经……。

[2] 长句，which 引导的是定语从句。instancing 动名词，意为 "示例"。

[3] 本句句型较复杂，that 引出修饰 shapes 的定语从句，…and so…引出结果从句，which 引出的是修饰 "Painting software" 的定语从句。

[4] 本句中使用了两个过去分词 required 和 not required 作定语，分别修饰 enhancements 和 features，由于过去分词作定语时有被动的意味，由 by 引出行为主体；who 引导的定语从句修饰 user。

[5] 长句，主句是 A template…is，中间插入介词短语构成的状语；that 引导的是定语从句，破折号后面是同位语，cut in it 为过去分词短语作定语，修饰 letters，it 代表 plastic sheet；which 引导的定语从句，修饰 plastic sheet；trace 为描绘，描摹。

[6] it 代表 template，dummy 为虚的，伪的，one 代表 "dummy" text。

[7] that 引出表语从句；in creating… in choosing 是同位关系，so 引导的是目的状语从句。

KEYWORDS

bit-mapped image	位图图像
pattern	模式
vector	矢量，向量
template	模板
dummy	虚化
CAD (Computer-Aided Design)	计算机辅助设计
graphic workstation	图形工作站
2D (Dimension)-plan	二维平面
3D (Dimension) object	三维物体
illustration software	绘图软件
library of drawings	图形库
business presentation software	商业演示软件
pattern	图案，模式

CHAPTER 10 COMPUTER GRAPHICS AND IMAGES

EXERCISES

1. Fill in the blanks with appropriate words or phrases found behind this exercises.

(1) One of the simplest types of CAD is the preparation of _____ plans.

(2) Though CAD software is associated with engineering design and other forms of technical drawing, the facilities offered by such packages have proved useful in a wide variety of other_____.

(3) The differences between CAD and _____ lie in emphasis and direction rather than principle.

(4) In graphics software a _____ is a design which the user can quickly adapt to their own purpose.

(5) _____ is a vector software with a very focused application.

(6) _____ has now spread well beyond the boundaries of CAD.

    a. applications   b. 2D

    c. template   d. illustration software

    e. Business presentation software   f. 3-D modeling

2. True/False.

(1) _____ CAD work is only done on graphic workstation.

(2) _____ The more powerful CAD packages can be used to design and represent 3D objects.

(3) _____ Illustration (or simply Drawing) software can be thought of as CAD software with a less specialized purpose.

(4) _____ A template, in the ordinary use of the term, is a pattern or guide that is used to help in the creation of some object.

(5) _____ Only can Illustration software be used to assemble and edit a range of line-based geometric shapes.

(6) _____ A presentation should include informative talks or short articles.

CHAPTER 11
MODERN INDUSTRIAL AUTOMATION

11.1 USE OF CAD, CAM, AND CAE

We have described a typical product cycle. Now we will review it to show how the computer, or CAD, CAM, and CAE technologies, are employed in the cycle. As indicated earlier, the computer is not widely used in the synthesis phase of the design process because the computer does not handle qualitative information well. However, in the synthesis subprocess, for example, a designer might well collect the relevant design information for the feasibility study by using a commercial database and collect catalog information in the same way.

Nor is it easy to imagine how a computer might be used in the design conceptualization phase because the computer is not yet a powerful tool for the intellectual creative process[1]. The computer may contribute in this phase by physically generating various conceptual designs efficiently[2]. The parametric modeling or macroprogramming capability of computer-aided drafting or geometric modeling may be useful for this task. These packages are typical examples of CAD software. You may imagine a geometric modeling system to be a three-dimensional equivalent of a drafting system; that is, it is a software package by which a three-dimensional shape instead of a two-dimensional picture is manipulated[3].

The analysis subprocess of the design process is the area where the computer reveals its value. In fact, there are many available software packages for stress analysis, interference checking, and kinematic analysis. These software packages are classified as CAE. One problem with using them is the provision of the analysis model. It would not be a problem at all if the analysis model were derived automatically from the conceptual design. However, as explained previously, the analysis model is not the same as the conceptual design but is derived by eliminating unnecessary details from the design or by reducing its dimensions. The proper level of abstraction differs, depending on the type of analysis and the desired accuracy of the solution. Thus it is difficult to automate this abstraction process; accordingly the analysis model is often created separately. It is a common practice to create the abstract shape of the design redundantly by using a computer-aided drafting system or a geometric modeling system or sometimes by using the built-in capability of the analysis packages[4].

The analysis subprocess can be embedded in the optimization iteration to yield the optimal design. Various algorithms for finding the optimal solution have been developed, and many optimization procedures are commercially available. Optimization procedures could be thought of as a component of CAD software, but it is more natural to treat optimization procedures separately.

The design evaluation phase also can be facilitated by use of the computer. If we need a design prototype for the design evaluation, we can construct a prototype of the given design by using software packages that automatically generate the program that drives the rapid prototyping machine[5]. These packages are classified as CAM software. Of course, the shape of the prototype to be made should exist in advance in a type of data. The data corresponding to the shape are created by geometric modeling. Even though the prototype can be constructed conveniently with rapid prototyping, it would be even better if we could use a virtual prototype, often called digital mock-up, which provides the same valuable information[6].

As the analysis tools used to evaluate the digital mock-up become powerful enough to give an analysis result as accurate as that from the equivalent experiment on a real prototype, digital mock-ups will tend to replace real prototypes[7]. This tendency will increase as virtual reality technology enables us to get the same feeling from the digital mock-up as we get from the real prototype[8]. The activity of building digital mock-ups is called virtual prototyping. The virtual prototype can also be generated by a kind of geometric modeling that is specialized for that purpose.

The final phase of the design process is design documentation. In this phase, computer-aided drafting is a powerful tool. The file-handling capability of computer drafting systems also allows the systematic storage and retrieval of documents.

Computer technologies are also used in the manufacturing process. The manufacturing process includes the activities of production planning, design and procurement of new tools, ordering materials, NC programming, quality control, and packaging, so all the computer technologies for these activities can be classified as CAM. For example, computer-aided process planning (CAPP) software to aid the process planning activity is one type of CAM software. As mentioned previously, process planning is difficult to automate, and thus 100 percent automatic CAPP software is not available currently. However, there are many good software packages that generate the numerically controlled (NC) programs that drive NC machines. This type of machine creates a given shape when the shape exists in the computer in the form of data. This is similar to driving the rapid prototyping machine. In addition, also belonging to CAM are the software packages to program robot motion to assemble components or deliver them to the various manufacturing activities, or to program a coordinate measuring machine (CMM) to inspect the product[9].

NOTES

[1] 将否定词 nor 提至句首强调否定意义时，要求主谓倒装；because 引导一个原因状语从句；how 引导一个宾语从句。

[2] in this phrase，在这一阶段；conceptual designs，概念设计。

[3] that is 为插入语，it 代表 geometric modeling system；by which 引导的是定语从句，修饰 software package。

[4] It 为先行代词，实际的主语是 to create…；abstract shape 可译为抽象模型；by using…一直到句末为两个分词短语作状语。

[5] If 引导一个条件状语从句，we can…为主句。by using…为方式状语，第一个 that 引导定语从句修饰 packages，第二个 that 引导定语从句修饰 program。

[6] Even though 引导让步状语从句，it would be…为主句，if…为条件状语从句。often called…为过去分词短语作定语修饰 prototype，which 引导定语从句亦修饰 prototype；mock-up 为模型。

[7] As the analysis…是一个状语从句，其中又包含由 as…as…引导的同等比较状语从句。

[8] 第一个 as 引导的是时间状语从句；第二个 as 引导的是方式状语从句。

[9] 倒装句，既强调 also belonging to CAM，又使长主语便于展开。正常语句应该是 the software package…are belonging to CAM。

KEYWORDS

interference checking	干扰检验
kinematic analysis	运动分析
iteration	迭代
virtual prototype	虚拟原型
NC (Numerical Control)	数控
CAPP (Computer-Aided Process Planning)	计算机辅助工艺计划
CMM (Coordinate Measuring Machine)	坐标测量机
commercial database	商业数据库
macroprogramming	宏编程，宏程序设计
imbed(embed)	嵌入
optimal design	优化设计
type of data	数据类型
retrieval	检索
robot	机器人

EXERCISES

Fill in the blanks with appropriate words or abbreviations found behind this exercises.

1. The computer is not widely used in the synthesis phase of the design process because the computer does not handle _____ information well.

2. Nor is it easy to imagine how a computer might be used in the design conceptualization phase because the computer is not yet a powerful tool for the _____ creative process.

3. It would not be a problem at all if the analysis _____ were derived automatically from the conceptual design.
4. The analysis subprocess can be embedded in the optimization iteration to yield the _____.
5. If we need a design prototype for the design evaluation, we can construct a _____ of the given design by using software packages that automatically generate the program that drives the rapid prototyping machine.
6. The final phase of the design process is design _____.
7. Computer-aided process planning software to aid the process planning activity is one type of _____ software.
8. Process planning is difficult to automate, and thus 100 percent automatic _____ software is not available currently.
9. Depending on the type of analysis and the desired accuracy of the solution, the proper level of _____ differs.
10. A software package that programs _____ to inspect a product is belonging to CAM.

    a. CAPP   b. abstraction
    c. documentation   d. qualitative
    e. CMM   f. CAM
    g. prototype   h. model
    i. intellectual   j. optimal design

11.2 3D PRINTING

3D printing (or additive manufacturing, AM) is any of various processes used to make a three-dimensional object. In 3D printing, additive processes are used, in which successive layers of material are laid down under computer control[1]. These objects can be of almost any shape or geometry, and are produced from a 3D model or other electronic data source. A 3D printer is a type of industrial robot. Figure 11-1 is an ORDbot Quantum 3D printer [2].

Figure 11-1 An ORDbot Quantum 3D printer[3]

3D printing in the term's original sense refers to processes that sequentially deposit material onto a powder bed with inkjet printer heads[3]. More recently the meaning of the term has expanded to encompass a wider variety of techniques such as extrusion and sintering based processes. Technical standards generally use the term additive manufacturing for this broader sense.

1. General principles

(1) Modeling

3D printable models may be created with a computer aided design (CAD) package or via a 3D scanner or via a plain digital camera and photogrammetry software.

The manual modeling process of preparing geometric data for 3D computer graphics is similar to plastic arts such as sculpting. 3D scanning is a process of analyzing and collecting digital data on the shape and appearance of a real object. Based on this data, three-dimensional models of the scanned object can then be produced.

Regardless of the 3D modeling software used, the 3D model (often in .skp, .3ds or some other format) then needs to be converted to either a .STL or a .OBJ format, to allow the printing (a.k.a. "CAM") software to be able to read it[4].

Figure 11-2 is a 3D model slicing.

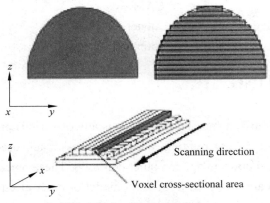

Figure 11-2　3D model slicing

(2) Printing

Before printing a 3D model from an STL file, it must first be examined for "manifold errors," this step being called the "fixup"[5]. Especially STLs that have been produced from a model obtained through 3D scanning often have many manifold errors in them that need to be fixed. Examples of manifold errors are surfaces that do not connect or gaps in the models[6]. Examples of software that can be used to fix these errors are netfabb and Meshmixer, or even Cura, or Slic3r [7].

Once that's done, the .STL file needs to be processed by a piece of software called a "slicer" which converts the model into a series of thin layers and produces a G-code file containing instructions tailored to a specific type of 3D printer (FDM printers) [8]. This G-code file can then be printed with 3D printing client software (which loads the G-code, and uses it to instruct the 3D printer during the 3D printing process). It should be noted here that in practice the client software and the slicer are often combined into one software program[9]. Several open source slicer programs exist, including Skeinforge, Slic3r, and Cura-engine as well as closed source programs including Simplify3D and KISSlicer[10]. Examples of 3D printing clients include Repetier-Host, ReplicatorG, Printrun and Cura[11].

(3) Finishing

Though the printer-produced resolution is sufficient for many applications, printing a slightly oversized version of the desired object in standard resolution and then removing material with a higher-resolution subtractive process can achieve greater precision[12].

Some printable polymers allow the surface finish to be smoothed and improved using chemical vapour processes.

Some additive manufacturing techniques are capable of using multiple materials in the course of constructing parts. These techniques are able to print in multiple colors and color combinations simultaneously, and would not necessarily require painting.

All of the commercialized metal 3-D printers involve cutting the metal component off of the metal substrate after deposition. A new process for the GMAW 3-D printing allows for substrate surface modifications to remove aluminum components manually with a hammer[13].

2. Applications

AM technologies found applications starting in the 1980s in product development, data visualization, rapid prototyping, and specialized manufacturing. Their expansion into production (job production, mass production, and distributed manufacturing) has been under development in the decades since. Industrial production roles within the metalworking industries achieved significant scale for the first time in the early 2010s. Since the start of the 21st century there has been a large growth in the sales of AM machines, and their price has dropped substantially. According to Wohlers Associates, a consultancy, the market for 3D printers and services was worth $2.2billion worldwide in 2012, up 29% from 2011. There are many applications for AM technologies, including architecture, construction, industrial design, automotive, aerospace, military, engineering, dental and medical industries, biotech (human tissue replacement), fashion, footwear, jewelry, eyewear, education, geographic information systems, food, and many other fields[14].

Figure 11-3 Model of a turbine showing the benefits of 3D printing in industry.

In 2005, a rapidly expanding hobbyist and home-use market were established with the inauguration of the open-source RepRap[15]. Virtually all home-use 3D printers released to-date have their technical roots in the ongoing RepRap Project and associated open-source

Figure 11-3　Model of a turbine showing the benefits of 3d printing in industry

software initiatives[16]. In distributed manufacturing, one study has found that 3D printing could become a mass market product enabling consumers to save money associated with purchasing common household objects[17]. For example, instead of going to a store to buy an object made in a factory by injection molding (such as a measuring cup or a funnel), a person might instead print it at home from a downloaded 3D model[18].

NOTES

[1] laid down，为建造，铺设之意，laid 为 lay 的过去分词，which 代表 additive processes。

[2] ORDbot Quantum 3D printer，ORDbot 量子 3D 打印机。

[3] that 引导的是定语从句，修饰 processes。

[4] 长句，本段就是一句话，句中 to allow…是目的状语，to be able…是动词不定式作复合宾语。a.k.a. 即 also known as。.skp 通常指的是 SketchUp 这个软件，中文名为"草图大师"，这个软件在建筑设计中越来越多地被使用，它和传统的 3D 软件相比有可视性强、易操作等特点，在设计初期对建筑形体关系进行推敲的时候经常用到。

STL 文件是在计算机图形应用系统中用于表示三角形网格的一种文件格式。它的文件格式非常简单，应用很广泛。OBJ 是一种 3D 模型文件格式，由 Alias|Wavefront 公司为 3D 建模和动画软件 Advanced Visualizer 开发的一种标准，适合用于 3D 软件模型之间的互导，也可以通过 Maya 读写。3ds 即 3D Studio Max，常简称为 3ds Max 或 MAX，是 Discreet 公司开发的（后被 Autodesk 公司合并）基于 PC 系统的 3D 动画渲染和制作软件。其前身是基于 DOS 操作系统的 3D Studio 系列软件。CAM（Computer Aided Manufacturing）即计算机辅助制造，是利用计算机来进行生产设备管理控制和操作的过程。

[5] it 代表 3D model；fixup，准备，安排，选定，整顿。

[6] connect 此处是指 3D 打印连接，gaps 是指 3D 打印空白，此处系指补缺。

[7] netfabb 是一款常用的模型编辑软件，可以测量、修复以及检查模型文件。我们熟悉的 Shapeways，也是使用 netfabb 提供的服务，对用户上传的模型文件进行检查。

Cura 提供 3D 打印模型使用专门的硬件设备的简单方法。Slic3r 是一款用于将 STL 文件转换成 GCode 的开源软件，它具有更加快速生成，可配置参数更加灵活等特点。Meshmixer 是一款 Autodesk 开放的 3D 模型设计软件，主要用于修改、拼接模型，还可以对模型进行结构分析，添加支撑，以备打印。

[8] 长句，在 which 引导的定语从句中，有 converts 和 produces 两个谓语，第二个谓语后面的宾语 a G-code file 又先后有现在分词 containing 和过去分词 tailored 短语作定语，分别修饰其前面的名词。

slicer 此处是指 3D Slicer，是一个全面的、有效的开源软件包。G-code（易录宝）是一种自动录影制式。FDM（Fused Deposition Modeling）即熔融沉积成型工艺，熔融沉积制造工艺。

[9] that 引导的是主语从句。

[10] closed source program，封闭源程序。

Skeinforge 是由 Enrique Perez 独立编写，是完整的工具链系统，能把 3D 模型转换为命令打印机的机器语言 G-code。Cura 是 Ultimaker 公司设计的 3D 打印软件，使用 Python 开发，并集成 C++开发的。Cura-engine 作为切片引擎，由于其切片速度快，切片稳定，对 3D 模型结构包容性强，设置参数少等诸多优点，拥有越来越多的用户群。

Simplify3D 是 2014 年 7 月 9 日德国 3D 打印公司 German RepRap 推出的一款全功能（All-in-One）3D 打印软件。Simplify3D 可取代 Repetier-Host 和 Slic3r 软件，支持导入不同类型的文件，可缩放 3D 模型，修复模型代码，创建 G 代码并管理 3D 打印过程。

KISSlicer 的全名是 Keep It Simple Slicer，意指"保持简单"的切片软件。

[11] ReplicatorG 的主要功能是把你的模型转换为打印机能识别的格式，但也可以用它对模型做一些基本的修改，使打印效果更好。Printrun 这款软件不仅有机器控制功能，还能和切片软件整合为一体（例如 slic3r），因此它可以独立完成从切片到打印的整个过程。它支持 Mac、Linux 和 PC 操作平台，几乎所有的开源 3D 打印机都可以使用这款软件。Repetier Host 是一款操作简单，将生成 Gcode 以及打印机操作界面集成到一起的软件，另外可以调用外部生成 Gcode 的配置文件，很适合初学者使用，尤其是手动控制的操作界面，可以很方便地实时控制打印机。

[12] Though 引导的是让步状语从句，主句是 and 连接的并列句。subtractive process，为 3D 打印中的减法处理。

[13] GMAW 3-D printing，熔化极气体保护焊 3D 打印。

[14] 长句，including…到末尾，列举了许多 AM 技术的应用。

[15] RepRap，是一个 3D 打印机原型机（或 3D 打印机），它具有一定程度的自我复制能力，能够打印出大部分自身的（塑料）组件。RepRap 是（replicating rapid prototyper）的缩写。

[16] technical roots，技术根基，技术基础。

[17] that 引导的是宾语从句，enabling…分词短语作定语。

[18] injection molding，注入成型。句中 instead of 作介词用，意为代替……，后面一个 instead 为副词，意为更换成……。

KEYWORDS

3D (3 Dimension)	三维
AM(Additive Manufacturing)	增材制造
process	过程，进程，处理
object	目标，对象，客体
successive layers	相邻层，连续层
shape	形状，整形
robot	机器人
deposit	沉积，沉淀
inkjet	喷墨
extrusion	挤压，伸长
modeling	建模，模型化，造型，模拟
CAD(Computer Aided Design)	计算机辅助设计
scanner	扫描仪
plain digital camera	普通数码相机
photogrammetry	摄影测量
plastic art	造型美术
sculpt	雕刻，雕像
manifold error	复印错误，印刷错误
clicer	限幅器，切片机
client	客户
closed source	封闭源
resolution	分辨率，清晰度，分解
precision	精度，精确，精密
prototyping	原型法
visualization	可视化，目视，显像
architecture	体系结构，结构，层次结构，总体结构
construction	构造

EXERCISES

Fill in the blanks with appropriate words or phrases found behind this exercises.

1. 3D printing uses a _____.
2. In 3D printing _____ of material are laid down under computer control.
3. In 3D printing objects are produced from a _____.
4. A 3D printer is a type of _____.
5. 3D printable models may be created with _____.
6. In order to prepare geometric data for 3D computer graphics, we need the _____.
7. _____ is a process of analyzing and collecting digital data on the shape and appearance of a real object .

CHAPTER 11 MODERN INDUSTRIAL AUTOMATION

8. CAM stands for _____.
9. The 3D model is often in _____ format.
10. Examining manifold errors is called as the _____.
11. Using _____ format can cause many manifold errors.
12. A _____ is a software that converts a model into a series of thin layers.
13. _____ file contains instructions tailored to a specific type of 3D printer.
14. In practice _____ and the slicer are often combined into one software program.
15. Cura-engine is a _____ slicer program.
16. The printer-produced resolution is sufficient for _____.
17. _____ can be used to smooth the surface of some printable polymers.
18. Some additive manufacturing techniques are able to print in _____ simultaneously.
19. All of the _____ involve cutting the metal components off the metal substrate after deposition.
20. Since the start of the 21st century the sales of _____ have been a large growth.
21. Virtually all home-use 3D printers have their _____ in the ongoing RepRap Project.
22. We can make a measuring cup at home by using a 3D model _____.
    a. manual modeling
    b. G-code
    c. AM machines
    d. fixup
    e. many applications
    f. downloaded from Internet
    g. additive processes
    h. .skp or .3ds
    i. 3D printing client software
    j. Chemical vapour processes
    k. 3D scanning
    l. 3D model
    m. technical roots
    n. Computer Aided Manufacturing
    o. CAD
    p. slicer
    q. commercialized metal 3D printers
    r. open source
    s. STL
    t. industrial robot
    u. multiple colors and color combinations
    v. successive layers

ANSWERS TO THE EXERCISES

1.1	1. f	2. h	3. o	4. n	5. i	
	6. m	7. d	8. p	9. g	10. a	
	11. l	12. e	13. k	14. b	15. c	16. j
1.2	1. a, b, c, d	2. a, b, d	3. b, c, d	4. a, c, d	5. a, b, c, d	
	6. b, c, d	7. b, d	8. a, b, c, d	9. b, c, d	10. b, d	
	11. a, c, d	12. a, b, c, d				
1.3	1. f	2. m	3. i	4. a	5. h	
	6. l	7. r	8. b	9. g	10. o	
	11. p	12. e	13. k	14. d	15. q	
	16. j	17. n	18. c			
1.4	1. f	2. t	3. t	4. f	5. t	
	6. t	7. f	8. f	9. t	10. t	
1.5	1. c	2. e	3. g	4. b	5. h	6. j
	7. a	8. i	9. f	10. d		
2.1	1. b, d	2. a, b, c, d	3. b, c, d	4. a, c	5. a, b, d	6. a, b, c, d
	7. a, b, c	8. a, b, c, d	9. a, b, d	10. b, d	11. c, d	12. a, b, c, d
2.2	1. d	2. l	3. g	4. k	5. e	6. b
	7. a	8. j	9. h	10. c	11. i	12. f
2.3	1. m	2. n	3. k	4. j	5. i	
	6. b	7. a	8. e	9. h	10. d	
	11. l	12. g	13. c	14. f		
3.1.1	1. g	2. c	3. l	4. d	5. k	6. f
	7. j	8. a	9. h	10. e	11. i	12. b
3.1.2	1. a, b, c, d	2. a, b	3. b, c, d	4. a, b, c, d	5. b, c	
	6. a, d	7. a, b, c, d	8. a, b, c	9. b, c, d	10. a, b, c	
	11. a, c, d	12. a, b, c, d				
3.2	1. a, b, c	2. a, b, d	3. b, c	4. a, c	5. b, d	
	6. b, c, d	7. a, c, d	8. b, c	9. a, b, c, d	10. b, c, d	
	11. a, b, c, d	12. a, c, d	13. a, b, c, d	14. b, c, d		
4.1.1	1. a, c, d	2. b, c, d	3. a, b, d	4. a, b, c	5. a, b, c, d	
	6. b, c, d	7. a, b, c, d	8. a, c, d	9. b, c, d	10. a, c	
4.1.2	1. f	2. f	3. t	4. f	5. f	
	6. t	7. t	8. f	9. t	10. f	
	11. t	12. t				
4.2	1. c	2. i	3. f	4. m	5. b	

ANSWERS TO THE EXERCISES

	6. j	7. d	8. n	9. g	10. a	
	11. l	12. h	13. k	14. e		
4.3.1	1. a, b, c, d	2. b, c, d	3. a, c, d	4. a, b, c, d	5. a, b, c, d	
	6. a, b, d	7. a, b, c, d	8. a, c, d			
4.3.2	1. t	2. f	3. t	4. t	5. f	
	6. t	7. f	8. t	9. f	10. t	
	11. f	12. t	13. f	14. t	15. t	16. t
5.1.1	1. b, d	2. a, b, c, d	3. a, b, c, d	4. c, d	5. a, d	
	6. a, b, c	7. c, d	8. a, c, d	9. a, b, c, d	10. a, b, c, d	
5.1.2	1. t	2. t	3. f	4. t	5. f	
	6. t	7. t	8. f	9. f	10. t	
	11. t	12. f	13. t	14. t		
5.1.3	1. c	2. g	3. e	4. i	5. h	
	6. d	7. j	8. b	9. a	10. f	
5.1.4	1. c	2. g	3. l	4. j	5. b	6. i
	7. f	8. a	9. e	10. h	11. k	12. d
5.2.1	1. t	2. t	3. t	4. f	5. t	
	6. f	7. t	8. t	9. f	10. t	
	11. f	12. t	13. t	14. f	15. t	
	16. t	17. f	18. t			
5.2.2	1. a, b, c, d	2. a, b, d	3. b, c, d	4. a, c, d	5. a, b, c, d	
	6. a, b, c	7. a, b, d	8. b, c, d	9. a, b, c, d	10. a, b, c, d	
	11. a, b, c, d	12. b, c, d				
5.3	1. a, b, d	2. a, b, c, d	3. b, c	4. a, b, c, d	5. b, c, d	
	6. a, b, c, d	7. a, c, d	8. b, c	9. b, c, d	10. a, b, c, d	
	11. a, c, d	12. a, c	13. b, c, d	14. a, b, d	15. a, b, c, d	
	16. a, c, d	17. a, b	18. a, b, c, d			
5.4	1. t	2. f	3. t	4. t	5. f	
	6. t	7. t	8. f	9. t	10. f	
	11. f	12. t	13. t	14. t	15. f	
	16. t	17. t	18. t	19. f	20. f	
6.1	1. a, b, c, d	2. a, b, c, d	3. b, c, d	4. a, b, c	5. a, b, d	
	6. a, b, c, d	7. a, c, d	8. b, c	9. a, c, d	10. a, b, c, d	
6.2	1. c	2. f	3. n	4. k	5. a	6. p
	7. j	8. o	9. g	10. l	11. b	12. r
	13. i	14. q	15. d	16. h	17. m	18. e
6.3	1. f	2. t	3. f	4. t	5. f	
	6. t	7. t	8. t	9. t	10. f	

	11. f	12. t	13. f	14. t	15. t	
	16. t	17. f	18. f	19. t	20. t	
7.1	1. a, b, c, d	2. a, b, d	3. a, b, c, d	4. b, c, d	5. a, b, c, d	
	6. a, b, c, d	7. a, b, d	8. a, b, c, d	9. b, c, d	10. a, b, d	
	11. a, b, c, d	12. b, c, d				
7.2	1. t	2. t	3. t	4. t	5. f	
	6. t	7. f	8. f	9. t	10. t	
8.1	1. a, b, c, d	2. a, b, d	3. a, b, c, d	4. b, c, d	5. a, c, d	6. a, b, d
	7. b, c	8. a, b, d	9. a, b, d	10. a, c, d	11. a, b, c, d	12. a, b, c, d
8.2	1. d	2. j	3. g	4. o	5. a	
	6. m	7. i	8. e	9. n	10. c	
	11. l	12. h	13. p	14. b	15. k	16. f
9.1	1. b, d	2. a, b, c, d	3. a, b, d	4. b, c, d	5. a, b, c, d	
	6. b, c, d	7. a, c, d	8. a, b, d	9. c, d	10. a, b, c, d	
9.2	1. f	2. c	3. i	4. e	5. a	6. l
	7. b	8. d	9. g	10. k	11. j	12. h
10.1	1. c	2. d	3. b	4. f	5. a	
	6. e	7. g				
10.2	1. (1) b	(2) c	(3) a	(4) a	(5) b	
	(6) b	(7) c	(8) b			
	2. (1) t	(2) t	(3) f	(4) t	(5) f	
	(6) t	(7) t	(8) f			
10.3	1. (1) b	(2) a	(3) d	(4) c	(5) e	(6) f
	2. (1) f	(2) t	(3) t	(4) t	(5) f	(6) t
11.1	1. d	2. i	3. h	4. j	5. g	
	6. c	7. f	8. a	9. b	10. e	
11.2	1. g	2. v	3. l	4. t	5. o	
	6. a	7. k	8. n	9. h	10. d	
	11. s	12. p	13. b	14. i	15. r	
	16. e	17. j	18. u	19. q	20. c	
	21. m	22. f				

参 考 译 文

第一部分　计算机体系结构和计算机网络

第1章　计算机组成和部件

1.1　计算机组成

1. 计算机的组成

计算机是一种可编程的电子设备，它接收输入、完成对数据的运算或处理，并输出和存储结果。因为它是可编程的，这些被称为程序的指令告诉计算机去做什么。计算机的这4种主要操作（输入、处理、输出和存储）之间的关系如图1-1所示。

完成这些任务的相应设备是输入设备、处理器、输出设备和存储器。

（1）输入设备

输入设备是任一种给计算机提供资料的设备。最常用的输入设备是键盘和鼠标，如图1-2所示。其他的有图像和条形码扫描器、操作杆、触摸屏、数码相机、电子笔、指纹阅读器和麦克风等。立体音响系统的输入设备是CD播放器和天线。

（2）处理器

任何计算机系统的核心都是中央处理器（CPU），放在计算机主机箱或系统单元中。

处理器由两个功能部件（控制部件和算术逻辑部件）与一组称为寄存器的特殊工作区组成。

图1-3描述了处理器的结构，其中，CPU的内部互连机构提供了控制部件、算术逻辑部件和寄存器之间的通信。

控制部件是负责监控整个计算机系统操作的功能部件。

控制部件从存储器中取出指令，并确定这些指令的类型或对其进行译码。然后将每条指令都分解成一系列简单的小步骤或动作。这样，它就控制了整个计算机系统的步进操作。

算术逻辑部件（ALU）是为计算机提供逻辑及计算功能的部件。控制部件将数据送入算术逻辑部件，然后由算术逻辑部件完成指令所要求的某些算术或逻辑运算。

寄存器是处理器内的存储单元。控制部件中的寄存器用来跟踪正在运行的程序的总体状态。控制器寄存器存储诸如当前指令、下一条将要执行的指令地址以及该指令的操作数等信息。在算术逻辑部件中，寄存器存放要进行加、减、乘、除以及要比较的数据项。而其他寄存器则存放算术和逻辑运算的结果。

（3）输出设备

与输入设备类似，输出设备也是人与各类计算机系统之间进行解释和通信的设备。输出设备从 CPU 中取出机器代码形式的输出结果，然后将其转换成人们可读的形式（例如打印或显示报告）或另一处理周期的机器输入代码。

在个人计算机系统中，常用的输出设备是显示器和台式打印机。比较大型的计算机系统通常要配备更大、更快的打印机、多台在线工作站和磁带机等。

（4）存储设备

存储设备是计算机的一部分，主要是用于存储诸如指令、程序和数据信息的。

存储器有两种类型，一是内存储器（有时称作主存储器），另一种是二级存储器。主存储器置于系统单元内，那里还有 CPU 和其他部件。二级存储器包括存储介质和驱动器，我们将在本书 1.4 节中叙述。

任意一台台式计算机（不一定必须是 PC）的框图如图 1-4 所示。它有一个很大的主存储器，用于保存操作系统、应用程序和数据，以及一个用于连接海量存储设备（磁盘、数字视盘/光盘只读存储器）的接口。它有各式各样的输入输出设备，用户用来实现输入（键盘、鼠标和音频设备）、输出（显示接口和音响设备）以及连接功能（连网和外围设备）。快速处理器需要一个系统管理程序去监视它的内核温度和供电电压，并可进行系统复位。

2. 计算机的种类

计算机有 4 种类型：超级计算机、大型计算机、中型计算机和微型计算机。

（1）超级计算机是功能最强的计算机，这些机器的性能特别高，通常由大单位使用。IBM Blue Gene 被许多人视为是世界上最快的计算机。

（2）大型计算机安装在专门布线和有空调的房间内，尽管它的功能不如超级计算机，但它有很高的处理速度和很大的数据存储量。例如，保险公司用大型计算机去处理上百万投保人的信息。

（3）中型计算机，也称为服务器，是功能比大型计算机弱，但比微型计算机强的计算机。最初，这种计算机由中型公司或大型公司的部门使用，用于其业务需要。现今，中型计算机更多地服务于终端用户，用于从数据库检索数据，或访问应用软件等特殊需求。

（4）微型计算机，是功能最弱而使用最广和增长最快的计算机。有 4 种微型计算机类型：台式、笔记本、平板和手持计算机（见图 1-5）。台式计算机很小，可以放在桌面上或桌边，但搬动起来比较困难。笔记本计算机，又称膝上电脑，是很轻的、便携式的，可以放在公文包里。平板电脑，又称平板计算机，是最新型的计算机，它比笔记本计算机更小、更轻，耗电也少。与笔记本计算机一样，平板电脑也有一个平面显示屏，但一般没有标准键盘，它使用一个在显示屏上出现的虚拟键盘，并且是触摸式的。最出名的平板电脑是苹果公司的 iPad 。手持计算机是最小的，可以放在手掌上。这些系统都是一个完整的计算机系统，包括电子部件、二级存储器和输入输出设备。个人数字助理（PDAs）和智能电话是使用最多的手持电脑。智能电话是蜂窝式电话，采用无线方法连入因特网。过去几年间，这些设备数量剧增。

1.2 微处理器和主板

1. 微处理器

在微型计算机系统中，中央处理器（CPU）或称处理器，包含在一个芯片上，又称为微处理器。微处理器既可以安装在插入到主板上的承载插件上，也可以安装在插入到主板专用插槽上的盒上，如图1-6所示。微处理器是计算机系统的"大脑"。它有两个基本部件：控制器和算术-逻辑部件。这两个部件的功能已在本书的1.1节中介绍过。

2. 微处理器芯片

芯片容量通常用字长表示。字是CPU一次可以访问的位数（如16、32或64），一字的位数越多，计算机的功能就越强，速度就越快。正如前面所提到的，8位组成一字节。32位的计算机一次可以访问4字节。64位的计算机一次可访问8字节，因而能处理64位字的计算机是更快的。

早期的微机每秒钟能处理几百万个数据和指令，或称微秒级。比较新的微机处理数据和指令更快，为每秒几十亿个或称毫微秒级。相比之下，超级计算机运算速度以皮秒度量，速度比微型机快千倍以上，如图1-7所示。

微处理器两个最新最重要的发展是64比特处理器和双核芯片。直到最近，64比特处理器只用于大型和超级计算机中。而当64比特处理器在当今功能较强的微机中都在使用时，就表明一切都在变化之中。

这种新型双核芯片有两个分开且独立的CPU。这些芯片允许一台计算机在同一时间运行两道程序。例如，在端用户用PowerPoint去制作多媒体演示文稿时，另一程序可同时搜索一个大型数据库。而更重要的功能是以前只能在大型和超级计算机上运行的非常复杂的程序也可在微机上运行。这要求把程序专门设计成可以在每个CPU上单独处理的两部分。这种方法称为并行处理。

3. 多核处理器

多核处理器是有两个或多个独立的中央处理器（称为"核"）的单个计算机部件。图1-8为通用双核处理器框图。

制造厂商通常把几个核集成在一个集成电路模片上（称为单片多处理器-CMP），或集成在多个模片上，再封装成一个芯片盒。

多核处理器可以有2核（双核CPUs，例如AMD Phenom Ⅱ X2 and Intel Core Duo）、4核（4核CPUs，例如Intel's i7 processors）、6核、8核或更多的核。

多核处理器广泛用于多个领域，包括通用领域，嵌入式，网络，数字信号处理以及图形学领域。

商业上，Adapteva公司的Epiphany架构，是一个多核处理器体系，可以在一个芯片上有多达4096个处理器，尽管推向市场的只是16核的版本。

4. 主板

主板也叫母板。主板是整个计算机系统的通信媒体。系统部件中的每个部件都连向主板。它的作用就像数据通路一样，允许各个部件之间相互通信。外部设备，如键盘、鼠标和监视器没有主板也不能与系统部件通信。

台式计算机的主板是在系统部件底部或在一侧。它是一块大的电路板，上面布满各种电子部件，包括插座、插槽和总线，如图1-9所示。

（1）插座是称为芯片的小型专用电子部件的连接点。芯片是由蚀刻在像沙子一样称为硅材料的方形薄片上的电路板组成的。这些电路板可能比你的指尖还小。芯片也可称为硅片、半导体器件或集成电路。芯片安装在承载插件内（见图1-10）。这些插件可以直接插入主板的插座上或放在卡上，然后再将这些卡插到主板的插槽上。插座用于将主板与不同类型的芯片，包括微处理器和存储器芯片连接起来。

（2）插槽是专用卡或电路板的连接点。这些卡为计算机系统提供扩展功能。例如调制解调器卡插入主板的槽中，可以与因特网连接。

（3）连接线，又称为总线，是连接各个电子部件之间的通信通路，这些电子部件安装在主板上或附着在主板上。

笔记本计算机、平板电脑和手持PC的主板比台式机的小，但它们实现的功能和台式机主板一样。

1.3 存储器

1. 存储系统的需求

存储系统有以下3项需求。

（1）容量：无限大，对程序和数据集大小没有任何约束。

（2）速度：无限快，等待时间在现有存储器技术下是最短的。

（3）价格：每位的价格在可使用的技术中应最低。

很明显，由于这3个需求相互制约，所以很难全部满足。不过随着半导体和磁存储技术的发展，这些需求几近得到满足。

2. 存储器

存储器是存放数据、指令和信息的地方。与微处理器一样，存储器是安装在与主板连接的芯片上的。大家熟知的存储器芯片有3种类型：随机存储器（RAM）、只读存储器（ROM）和互补金属－氧化物半导体（CMOS）。

（1）随机存储器（RAM）

随机存储器芯片保存CPU正在处理的程序（指令序列）和数据。由于数据在处理之前，或程序在运行之前，它们必须先保存在RAM中，因而有时也称RAM为主存储器。RAM又称暂时或易失性存储器，这是由于大多数这类存储器当微机关闭时，其所存储的各种信息都会丢失。如果电源出故障或其他因素中断了对微机的供电，信息也会丢失。

随机存储器分为动态随机存储器（DRAM）和静态随机存储器（SRAM）两种类型。

① 动态随机存储器（DRAM）

最近生产的DRAM芯片有以下三种类型。

- 同步DRAM（SDRAM），比常规RAM芯片要快、要贵。SDRAM芯片采用流线型操作方式，方法是对该芯片和系统部件中的其他部件之间的数据和指令的传送进行协调同步。
- 双数据速率 SDRAM（DDR SDRAM），又称为 SDRAMII，它比 SDRAM 更快、更可靠、更贵。DDR SDRAM 芯片与 SDRAM 相比，在相同时间内传输数据的速率要快一倍。
- 直接 DRAM，是最快、最贵的一种芯片。

当今大多数微机都采用这几种 DRAM 芯片的组合方式。

② 静态随机存储器（SRAM）

静态 RAM 与 DRAM 一样，要求稳定的供电。与 DRAM 比较，SRAM 不需要太高的电源功率，但是更快更可靠。SRAM 也很贵，通常是专用的，这些应用之一是高速缓存或 RAM 缓存。

③ 高速缓冲存储器（缓存）

高速缓冲存储器改善了存储性能，其方式是将其作为存储器和 CPU 之间的一个暂时高速存储区。在有缓存的计算机中（并非所有计算机都有），计算机检测 RAM 中哪些信息是最常用的，并将其复制到缓存中，在需要时 CPU 能很快将这些信息从缓存中取出。

有三种不同类型或级别的缓存。

- 1 级（L_1），也称为主缓存和内部缓存，它放在微机处理器芯片上。
- 2 级（L_2），也称外部缓存，它比 1 级速率慢，但容量更大。在旧一点的微处理器构成的计算机中，2 级缓存是放在插入主板的芯片上的。由较新的微处理器构成的新型计算机有内置在微处理器上的 2 级缓存。这种安排有时是指先进的传输缓存，它比安装在主板上的缓存反应速度更快。
- 3 级（L_3），为最新的一种缓存，它与专用微处理器 L_2 缓存共同使用。L_3 缓存并未置入到微处理器中，而是使用主板上的同步动态 SDRAM。

当今大多数微机都有两种或三种缓存，功能最强的有全部三种缓存。

④ 闪存

闪存芯片可以在电源断开后仍然能保存数据。这种随机存储器是最贵的，主要用在专用设备中，如数字移动电话、数码相机和便携式计算机中。

（2）只读存储器（ROM）

只读存储器（ROM）芯片上有在工厂制造时写入的程序。与 RAM 芯片不同，ROM 芯片不是易失性的，用户不能更改它上面的内容。"只读"是指 CPU 只能读或检索写到 ROM 芯片上的数据和程序。但计算机不能写（即编码或改变）ROM 中的信息或指令。

通常 ROM 芯片上有用于计算机具体操作的专用指令。例如，ROM 中的指令可以启动计算机，为键盘上的键安排专用的控制功能，并将字符显示在屏幕上。

（3）互补金属氧化物半导体（CMOS）

互补金属氧化物半导体（CMOS）芯片能使计算机系统更灵活，更具可扩充性。它含有当计算机系统启动时所需的基本信息。该芯片提供当前日期和时间、RAM 容量、键盘、鼠标、监视器和磁盘驱动器的类型等信息。与 RAM 不同，CMOS 是由电池供电的，当电源关闭以后，其内容不会丢失。与 ROM 也不同，它上面的内容可以更改，以反映计算机系统有了变化，如增加了 RAM 和新的硬件设备等情况。

1.4 二级存储器

由于 RAM 是一种暂时或易失性存储器，因而更需要永久或非易失性存储器来存储数据和程序。又由于用户需要比通常的主存或 RAM 更大的存储容量，因此也需要外部存储器。

二级存储器是永久和非易失性存储器。使用像硬磁盘驱动器那样的二级存储器，数据和程序在计算机关闭之后仍得以保存。二级存储器的操作是通过向这些存储设备写入文件或从这些设备读出文件来实现的。

1．硬磁盘

与软磁盘相比，硬磁盘存储和检索信息更快，容量更大。

有三种类型的硬盘：

① 内部硬盘，内部硬盘放在系统部件内部。

② 盒式硬盘，盒式硬盘很容易取出，就像从录像机中取出录像带一样。

③ 硬盘组，硬盘组也是可移动存储设备，用于存储海量信息，其容量远远超过其他类型的硬盘。

有三种方法能改善硬盘性能，它们是磁盘高速缓存、独立磁盘冗余阵列（RAID）和文件压缩／解压缩。

2．光盘

现今的光盘可以存储大于 50 GB 的数据，相当于将几百万页打印稿或一个中等规模的图书馆都存在一张光盘上。

（1）高密度磁盘

高密度磁盘更应该称为光盘，是应用最广的光学形式之一。

光盘有三种基本类型：只读、写一次和可重写型。

① 只读 CD-ROM，是光盘只读存储器，类似于商用音乐光盘。只读表明用户不能去写或擦除。

② 写一次 CD-R，是一种可录制却只能写一次的光盘。此后可读多次而不损伤光盘，但不能再写和擦除。

③ 可重写 CD-RW，代表可重写光盘，也称为可擦除光盘。

（2）数字通用光盘（DVD）

数字通用光盘 DVD 又称为数字视频光盘。作为标准光盘，这是一种可以取代 CD 的较新的形式。

（3）高清晰光盘

高清晰光盘是下一代光盘，又称为 HiDef 光盘，容量远大于 DVD。与 CD 和 DVD 相同，HiDef 光盘也有三种基本类型：只读、写一次和可重写。

3．其他类型的二级存储器

（1）固态存储器

迄今为止所讨论的每一种二级存储器都有活动部分。例如，硬盘旋转、读／写头移进移出。与这些设备不同，固态存储器没有活动部分。数据和信息以电的形式直接存入这些设备或从这些设备检索出来，很像访问传统计算机存储器一样。

① 闪存卡，是信用卡大小的固态存储器，广泛用于笔记本计算机。

② USB 驱动器，又称 USB 闪存驱动器，这类驱动器很小巧，可以放在钥匙链或项链上携带。这些驱动器可以很方便地直接连入计算机 USB 端口，传送文件。

（2）因特网硬盘驱动器

在万维网专用服务网站上向用户提供存储空间，这种存储器称为因特网硬盘驱动器，如图 1-11 所示。

比起其他类型的二级存储器，因特网硬盘驱动器的优点是费用低，以及能从任何使用因特网的站点访问信息的灵活性。

（3）磁带机

磁带机是慢速顺序访问设备。虽然访问磁带上的信息速度低，但它仍是备份数据的有效和常用工具。

1.5 输入与输出设备

1. 输入设备

（1）键盘输入

输入数据最常用的方法之一是使用键盘。

键盘有多种不同的类型，从全尺寸的到小型的，从刚性的到柔性的，最常用的有：
- 传统键盘（见图 1-12）
- 柔性键盘（见图 1-13）
- 人体工程学键盘（见图 1-14）
- 无线键盘
- 个人数字助理（PDA）键盘

（2）定位设备

① 鼠标

鼠标控制在监视器上显示的指针。鼠标指针通常以箭头形式出现。虽然有多种不同的鼠标类型，但基本类型只有三种：机械鼠标、光学鼠标、无线鼠标。

② 操纵杆

操纵杆是计算机游戏中最常用的输入设备，如图 1-15 所示。

③ 触摸屏

触摸屏是一种特殊的监视器，通常用在饭店、自动取款机和信息中心等地方，如图 1-16 所示。

④ 光笔

光笔是由光敏元件做成的像笔一样的设备。使用时要将光笔对准监视器。这样它就闭合了光电子电路，因而确定了输入或修改数据的一个点。于是，光笔就可用于编辑数字图像和图形了。

⑤ 记录笔

记录笔类似一种笔，常与平板电脑和个人数字助理一起用，如图 1-17 所示。记录笔用压力在屏幕上画图。通过手写识别软件，记录笔与计算机交互。该软件把手写注记翻译成系统能处理的格式。

（3）扫描设备

扫描仪在文本和图像上面移动。扫描设备把扫描过的文本和图像转换成系统部件能处理的格式。有四种扫描设备：光学扫描仪、读卡器、条形码阅读器和字符与标记识别设备。

（4）图像捕获设备

图像捕获设备像传统复印机那样，是一种可对原件复印的光学扫描仪。

（5）音频输入设备

音频输入设备能把声音转换成系统部件可以处理的格式。至今，应用最广的音频输入设备是麦克风。

2. 输出设备

（1）监视器

最常用的输出设备是监视器，又称为显示屏。这种监视器展现的是文本和图形画面。这种输出常称作软拷贝。

（2）打印机

你大概经常会用打印机去打印布置的家庭作业、照片和网页。打印机把系统部件处理过的信息翻译好，然后在纸上打印出来。打印机输出通常称为硬拷贝。

3. 输入和输出组合设备

很多设备把输入和输出功能组合起来了。这样做有时是为了节省空间，有时是用在专用设备上。常用的组合设备包括传真机、多功能设备、因特网电话和终端。

第 2 章　系统软件

2.1　Windows 10

Windows 10 是微软作为 Windows NT 操作系统系列的一个成员开发并即将推出的操作系统。Windows 10，2014 年 4 月在 Build 会议上首次展现，并安排在 2015 年中期公布，而现在由 Windows Insider 程序进行公开的 beta 测试。在其提供使用的第一年内，对于合法的 Windows 7 和 Windows 8.1 用户，可以免费升级到 Windows 10。

Windows 10 的目标是围绕通用内核统一 Windows PC、Windows Phone、Windows Embedded 和 Xbox One 产品系列。这些产品都共享一个通用的、"万能"的应用架构和 Windows Store 生态系统，该生态系统扩展到由 Windows 8 引出的 Windows Runtime 平台。Windows 10 还将推出一种新的、绑定的万维网浏览器 Microsoft Edge，用它取代 IE 浏览器。

图 2-1 为 Windows 10 屏幕快照。

1．用户界面与桌面

Windows 10 改进了 Windows 8 的用户界面，这一界面按照使用的设备类型和所用的输入方法改变用户的习性。当一个键盘接入时，就会问用户是否希望切换到鼠标和键盘优化的用户界面模式，或停在触摸优化模式上。启用了一个新的重复开始菜单的方法，其方法是，在左侧有一个应用程序列表和"全部应用程序"按钮，在右侧有一个实时图形显示块。该菜单可以缩放，可以扩展成全屏幕显示，在触摸环境下，它是默认项。

一个新的称为任务查看的虚拟桌面系统加入进来。单击任务栏上的 Task View 按钮，或从屏幕左侧敲击，都会显示所有打开的窗口，并允许用户在这些窗口之间切换，或在多个工作空间之间切换。以前 Windows Store Apps 只能用在全屏幕方式，现在可用于桌面视窗方式或全屏幕方式。通过拖曳，可把程序视窗移到屏幕角上，只占屏幕的 1/4。当一个视窗被快速移动到屏幕的一侧时，用户就可以选择第二个视窗去填充未用的另一侧，这称为抓取助手。

2．特性

Windows 10 的主要特点是，在不同级别的设备之间，在 Windows 8 首次推出的 Windows 用户界面中出现的寻址不足问题上，集中协调用户的经验。

Windows Store app 生态系统被修订进"Windows apps"中。它们可以跨多个平台和设备级运行，包括智能电话、平板电脑、Xbox One 和其他与 Windows 10 兼容的设备。Windows apps 跨平台共享代码，有满足于设备要求和用于输入的响应式设计，能使 Windows 10 各设备之间的数据同步（包括通知，资历证明，并允许跨平台多人玩游戏），并通过统一的 Windows Store 进行分配。开发者也能"交叉购买"，这时购买的 app 许可证，可用于所有的用户兼容设备，而不是只能用于所购买的那个设备（即一个用户购买一个 PC 上的 app，也能在智能手机上使用，而无须额外付费）。

Windows 10 也允许 Web apps 和桌面软件（使用 Win32 或 .NET Framework）包装在一起，在 Windows Store 上销售。通过 Windows Store 销售的桌面软件将使用 App-V 系统包装，以允许沙箱工作。

3．版本

和以前一样，我们将提供适合于各种设备使用的不同 Windows 版本。这些版本满足不同客户的专门需求，从消费者到小型企业，到最大的企业。

- Windows 10 Home 是客户专用的桌面版，它为 PC、平板电脑和二合一电脑提供熟悉的个人使用环境。
- Windows 10 Mobile 用在像智能手机、平板电脑那样的小型、移动、触摸式设备上。
- Windows 10 Pro 是 PC、平板电脑和二合一的桌面版。它既有 Windows 10 Home 版相似的特性，也有改革之处。它有很多附加功能，能满足小型企业的各种需求。
- Windows 10 Enterprise 是构建在 Windows 10 Pro 上的，增加了先进功能，以满足中大型组织的需求。

- Windows 10 Education 是构建在 Windows 10 Enterprise 上的，以满足学校员工、管理人员、教师和学生的需求。
- Windows 10 Mobile Enterprise 是为企业客户的智能手机和小型平板电脑提供最佳手段的版本。

2.2 UNIX 和 Linux

1. UNIX

UNIX 操作系统最早是由美国电话电报公司（AT&T）贝尔实验室的丹尼斯·里奇和肯·汤普森开发的操作系统，允许计算机同时接纳多个用户并处理多道程序。从 20 世纪 70 年代 UNIX 开发以来，许多个人，特别是加利福尼亚大学伯克利分校的计算机科学家对它进行了改进（通常称为伯克利软件版本 UNIX 或 BSD UNIX）。这种操作系统在各类计算机系统，从个人计算机到大型机上广泛使用，并可以用其他相关形式使用。AIX 是运行在 IBM 工作站上的，A/UX 是在 Macintosh 计算机上运行的图形版本；Solaris 在英特尔微处理器上运行。

特性

（1）UNIX 系统能支持多用户和多任务

（2）UNIX 系统内核

内核是 UNIX 操作系统的心脏，负责控制计算机的资源和调度用户作业，以便每一个用户都能合理地共享资源。程序与内核的交互作用是通过带有熟知名字的专用函数实现的，这称为系统调用。

（3）外壳

外壳是一种命令解释程序，其作用如同用户与操作系统之间的接口。当你在终端上输入一个命令时，外壳解释该命令并调用你所要求的程序。

（4）设备无关的输入和输出

对于 UNIX 程序，设备（如打印机或终端）和磁盘文件都是作为文件出现的。当你给 UNIX 操作系统一个命令时，你可以指挥操作系统把输出送到任何设备或文件上去。这种变更称为输出重定向。

2. Linux

Linux 是一种类似 UNIX 的、基本上是遵循 POSIX 的计算机操作系统，是由免费和开源软件开发模型和 Linux 发行版汇集而成的。Linux 认定的部件是 Linux 核，是 1991 年 10 月 5 日由 Linux Torvalds 首先公布的操作系统的核。自由软件基金会（FSF）使用 GNU/Linux 这个名字去描述该操作系统，但引起了争论。

Linux 最初是为 Intel x86 型个人电脑开发的免费操作系统，但此后被移植到更多的硬件平台上，其数量超过任何其他操作系统。在服务器和其他大型机以及超级计算机中，Linux 是个领头的操作系统，但是用在台式机上仅占约 1.5%。 Linux 也运行在嵌入式系统中，在这些系统中，操作系统是典型的固化件，并与整个系统是高度融合的。安卓，这种在平板电脑和智能手机中应用最广的操作系统，就构建在 Linux 核的顶端。

（1）硬件支持

Linux 核是被广泛移植的操作系统核，它运行在多种计算机架构上，包括基于 ARM 的手持 iPAQ 和 IBM 大型系统 z9 或 z10 上——覆盖范围从手机到超级计算机，如图 2-2 所示。

（2）使用

除了在台式机和服务器上使用的通用 Linux 发行版外，还可以为不同的应用制作专用的发行版，这包括支持某种计算机体系结构，嵌入式系统，稳定性，安全，为某一地区或某种语言的本地化，目标指向一个专门用户组，支持实时应用或委托给一个给定桌面环境等。到 2015 年，已实际开发了 400 多种 Linux 发行版，其中大约 12 种是最普通的应用。

（3）桌面

Linux 在标准台式计算机和膝上电脑的普及程度与日俱增。目前最多的应用是带有两个最普通环境的图形用户环境：GNOME（它可以使用另外的外壳，如默认的 GNOME 外壳和 Ubuntu Unity），和 KDE Plasma Desktop。

并没有单个的正式的 Linux 桌面存在：需要从免费和开源软件池中根据桌面环境和 Linux 发行版去选择各种模块，用这些模块去构建一个具有或多或少精准设计指南功能的图形用户接口。例如，GNOME，作为一个设计指南，它有自己的人机接口指导原则，它给人机接口一个重要指导，而不只是做这个图形设计，它要考虑人的有限能力，甚至考虑到安全问题。

图 2-3 为 Linux 桌面栈可视化软件模块。

2.3 安卓操作系统

安卓是一种用于智能电话和平板电脑这样的触摸屏移动设备的操作系统，它是由 Open Handset Alliance 开发的，该组织是一个由谷歌公司领导的，由硬件、软件和通信公司组成的联合体。

安卓由基于 Linux kernel 的核构成，并配有中间件、库和用 C 语言写的应用程序接口，以及运行在包括基于 Apache Harmony 的、与 Java 兼容的库的应用框架上。

自 2013 年 7 月以来，Google Play 商店已经公布了 100 多万个安卓应用软件，下载应用软件已超过 500 亿个。一个开发者对 2013 年 4 月至 5 月间的调查表明，约 71%的移动开发者使用安卓进行开发。Google I/O 2014 会议上，谷歌公司透露，每个月的安卓用户都超过 10 个亿，而截至 2013 年 6 月，用户数为 5.38 亿。到 2015 年，安卓会成为最大的，所有通用操作系统使用的基础。

尽管大多数安卓设备最终都是以开源和专用软件的组合形式售出的，这包括由谷歌开发并授权的专用软件，但安卓的源代码是在开源许可下由谷歌公布的。

安卓对于那些需要成熟的、低价位的、客户化的，用于高技术设备的操作系统的技术公司来说，是很普通的。安卓的开放特性，促使大量开发者和热衷者，将这种开源码作为共同驱动课题的基础，这又会为资深用户增加新功能，或将安卓引入已经正式公布的运行其他操作系统的设备中。该操作系统的成功，又会使其成为各技术公司之间进行所谓"智能电话大战"一部分的专利诉讼的目标。

在当前的手持设备中，安卓的功能和技术规范包括：

（1）平台

该平台能适应较大型 VGA、二维图形库、基于 Open GL ES 2.0 规范的三维图形库和传统智能电话。

（2）存储

用于数据存储的是 SQLite，是一个小型关系型数据库。

（3）连通性

安卓支持的连通技术可连接 GSM / EDGE、IDEN、CDMA、EV-DO、UMTS、蓝牙和 Wi-Fi 等。

（4）消息

消息的可用格式有 SMS 和 MMS，包括线程的文本消息。现在安卓的云到设备消息（C2DM）也是 Android Push Messaging 服务的一部分。

（5）Web 浏览器

安卓里的 Web 浏览器是基于开源 WebKit 引擎的，是与 Chrome's V8 Java Script 引擎相配合的。

（6）Java 支持

虽然大多数安卓应用程序是用 Java 编写的，但在此平台上不使用 Java 虚拟机，也不执行 Java 字节代码。

（7）媒体支持

安卓支持音频 / 视频 / 静态媒体格式，包括 MP3、MIDI、WAV、JPEG、PNG、GIF、BMP 等。

（8）流媒体支持

支持 RTP / RTSP Streaming，HTML 连续下载，Adobe Flash Streaming（RTMP）和 HTTP Dynamic Streaming 等。

（9）附加硬件支持

安卓可以使用电视（Android TV）、汽车（Android Auto）、腕表（Android Wear）、静态/数码相机、游戏操作台、触摸屏和全球定位系统等。

（10）视频呼叫

安卓不支持本地视频呼叫，但某些手持设备有定制的、支持视频呼叫的操作系统版本。

另外一些支持手持设备的功能和规范有，多种语言、多任务、多触摸、蓝牙、基于音频的功能和屏幕捕捉等。

第 3 章　计算机网络

3.1　局域网

3.1.1　以太网

局域网是一种计算机网络，它在有限范围内，如家庭、学校、计算机房或办公楼内，把计算机用网络媒体相互连接起来。与广域网相比较，局域网的特点是它的较小的地域范围，以及不使用租用通信线路。

ARCNET、令牌环和其他技术标准是以前使用过的，但在双绞线上的以太网和 Wi-Fi 是目前构建局域网的两个最常用的技术。

网络拓扑叙述的是在设备和段之间相互连接的布局格式。在数据链路层和物理层，局域网有很多拓扑结构，包括环型、总线型、网型和星型。但现今最常用的局域网拓扑是交换式以太网。在高层，互联网协议(TCP/IP)已经成为标准，取代了 NetBEUI、IPX/SPX、AppleTalk 和其他协议。

简单的局域网一般由一个或几个交换机组成，一台交换机可以连接一台路由器、电缆调制解调器或 ADSL 调制解调器，以接入因特网。复杂的局域网由于采用具有生成树协议的交换机，且使用冗余链路，从而防止了回路效应，也具有通过服务质量管理不同类型流量的能力，也能用虚拟局域网技术去隔离流量。一个局域网可以包括很多网络设备，如交换机、防火墙、路由器、负载平衡器以及传感器。

局域网通过租用线路、租用设备或使用虚拟专用网技术的因特网，保持与其他局域网的连接。

1．网络交换机

网络交换机（又称为交换式集线器、桥式集线器，正式称为 MAC 桥）是在一个计算机网络内将设备连在一起的计算机网络设备（见图 3-1），是采用分组交换方式去接收、处理并向目的设备转发数据的。与落后的网络集线器不同的是，网络交换机只向一个或多个需要数据的设备转发数据，而不是向所有端口广播同一个数据。

2．10 千兆比特以太网

10 千兆比特以太网（10GE、10GbE 或 10GigE）是以 10Gb/s 速率（10×10^9 或每秒 100 亿位）传送以太网帧的一组计算机网络技术。这是首次由 IEEE 802.3ae—2002 标准定义的。与前面的以太网标准不同的是，10 千兆位以太网只规定使用由网络交换机连接的、全双工点对点链路；并未使用以前的共享媒体的 CSMA/CD 操作的以太网标准。半双工操作和集线器在 10GbE 中也不用了。

3．太比特以太网

脸谱网站和谷歌在众多公司中表示需要太比特以太网。有些人想，400 千兆比特比太比特（1000Gb/s）更实际。2011 年，研究人员预测，太比特以太网（1Tb/s）在 2015 年出现，而 100 太比特（100Tb/s）以太网会在 2020 年出现。

3.1.2　Wi-Fi 和蓝牙

1．Wi-Fi

Wi-Fi（通常用 WiFi，但不妥），是允许电子设备使用 2.4GHz UHF 和 5GHz SHF ISM 无线频带加入计算机网络的一种无线局域网技术。

Wi-Fi 联盟将 Wi-Fi 定义为任何"基于 IEEE 802.11 标准的无线局域网产品"。但是，术语"Wi-Fi"通常在英语中是"无线局域网"的同义语，这是因为大多数现在的无线局域网都是基于这些标准的。Wi-Fi 是 Wi-Fi 联盟的商标。这种"认证了的 Wi-Fi"商标只能由完全通过 Wi-Fi 联盟互操作性认证测试的 Wi-Fi 产品使用。

很多设备都可以使用 Wi-Fi，如个人电脑、视频游戏控制台、智能电话、数码相机、平板电脑和数字语音播放器。这些设备可以通过无线网络接入点接入像因特网那样的网络资源。这样的接入点（或称热点）室内的作用范围为 20 米（66 英尺），室外范围要大一些。热点覆盖形成的区域小可小到一间房间，其围墙阻挡了无线电波，大可大到几平方公里的范围，这时要使用重叠接入点技术。图 3-2 是笔记本计算机和打印机之间通过接入点进行通信的过程。

Wi-Fi 比起有线连接，如以太网，安全性要差一些，因为入侵无线网的人不需要连线。使用安全套接层（SSL）的网页是安全的，但未经加密的因特网访问，很容易被入侵者检测出来。鉴于此，Wi-Fi 采取了多种加密技术。早期的 WEP 加密很容易被攻破，后来加入了高质量协议（WPA，WPA2）。在 2007 年又加入了称为 Wi-Fi 保护设置（WPS）的选项，但它的严重瑕疵使得入侵者可以恢复路由器的密码。Wi-Fi 联盟后来修订了其测试计划和认证程序，以确保新的认证了的设备可以抵御攻击。

2．蓝牙

蓝牙是一种用于在固定和移动设备之间，短距离数据交换的无线技术标准（使用 ISM 波段中的短波 UHF 无线波段，范围为 2.4～2.485GHz），也可用来构建个人区域网。这项技术是由电信公司爱立信在 1994 年研发的。最初的设想是用无线技术替代 RS-232 数据电缆。它能连接几种设备，克服了一些同步问题。

蓝牙由蓝牙特殊兴趣组（SIG）管理，该组织的成员是电信、计算机、网络和消费电子学领域中的 25000 多家公司。IEEE 为蓝牙制订的标准是 IEEE 802.15.1，但它并未保持这个标准。蓝牙 SIG 关注着该规范的发展，管理认证过程，保护该商标。制造商所生产的设备必须遵守蓝牙 SIG 标准并标识其为蓝牙设备。

蓝牙是一个标准的取代连线的通信协议，主要用于低功耗、短距离环境下，在每台设备中，都有一个低价位的收发器微芯片。因为这些设备使用无线（广播）通信系统，它们相互之间不必具有可视视距，但准光学的无线通道必须是可见的。作用范围由功率决定，不过，有效范围还随实地情况而变，见图 3-3。

3．蓝牙与 Wi-Fi 对比

蓝牙与 Wi-Fi 有一些相似的应用：组建网络，打印或传输文件。Wi-Fi 作为高速电缆的替代物，通常用于局域网接入。这种类型的应用，有时也称为无线局域网。蓝牙主要用于手持设备。这类应用可视为在无线个人区域网中的应用。蓝牙在各种个人手持应用中，全部取代了电缆，并能在固定位置上应用，如在家庭中的智能能源应用（自动调温器等）。

Wi-Fi 和蓝牙在它们的应用中，在某些方面是互补的。Wi-Fi 的接入通常采用以点为中心的方式，具有非对称客户-服务器连接，所有通信量都通过该接入点路由，而蓝牙则在两个蓝牙设备之间采用对称方式。蓝牙在简单的应用中，如两个设备需要用最小的设置进行连接，如按下按键、头戴耳机和远程控制这些情况工作得很好，而 Wi-Fi 更适合于客户配置有一定档次，需要高速传输的应用场合，特别是通过接入节点接入网络的情况。

3.2 因特网

因特网是世界上最大的和最著名的计算机网。在技术上，因特网是一个网络的网络，这是因为单个用户都可以连接到由因特网接入提供商或因特网服务提供商（ISP）所建立的网络上，而该网络又可以连接到更大的网络上，而这个网络又连向了比它更大的网络上。所有这些网络的网络在一起就称为因特网。由于因特网上的所有网络是互联的，因此具有因特网访问功能的任何计算机都能在因特网上相互通信，而不必关心所使用的 ISP。

1. 域名系统

建立域名系统是为了集中管理把网络名变换成地址的任务，并使这种翻译功能自动化。早期的因特网，由中央单位（在加利福尼亚曼佬公园 SRI 网络信息中心斯坦福研究院）负责维护 HOSTS 文件，该文件包含了每一个因特网上的主机名连同它的地址。管理员必须把改变的内容传送给 SRI NIC，而且这些改变会定时合并到该文件中。当然，这意味着还必须把文件分发到每一个主机，以使它能有更新的版本。

DNS 采用一种遍布因特网的，跨越很多计算机的层次型分布式体系结构。根服务器保持有关顶级域的信息（像.COM、.EDU 和.GOV），并且整个因特网上的每个域都有一个域名服务器，负责该域中的计算机名和地址的对应。当客户计算机需要获得主机的地址时，它们查询 DNS 服务器。如果本地 DNS 知道该地址，它把该地址返回给客户机。如果它不知道，它把该查询送往 DNS 服务器链，直到查找到一个能分析这个名字的 DNS 服务器并提供一个真正有效的名字为止。

在 DNS 层次中，最顶级的项称为根域，并且它前面有一个句号（.）。在这个根域下面是顶级目录，它分成两组：地理域和组织机构域。地理域常用于指明国家。例如，.au 是指澳大利亚，.cn 是指中国。在每一个地理域下面，你能找到组织机构域。

组织机构域你可能熟悉，它包括下列各项：

（1）com　用于商业组织。
（2）edu　用于教育机构。
（3）gov　用于美国政府实体。
（4）mil　用于美国军事机构。
（5）Int　国际性组织。
（6）net　用于网络组织，像因特网服务提供者。

（7）org   用于非营利的组织。
（8）arpa   用于逆向地址查找。
域名系统的结构和一棵倒长着的树很相似。在图 3-4 中，你可以看到顶端是根域，在它下面有 com 到 cn 的各个域。在 com 域下面是单独的商业机构，他们都有自己的域。在任一具体的域下面，都可能有子域。

2．连入因特网
（1）拨号连接
拨号连接一般是在普通电话线上连接。为了连向因特网，你的调制解调器（或其他接口设备）要拨号并与你的 ISP 计算机上的调制解调器相连。在你连接时，要对此次会话为你的 PC 安排一个临时 IP 地址。在每次因特网会话结束时，要与你的 ISP 计算机断开，以便其他用户使用该地址进行连接。标准因特网拨号业务使用常规的拨号调制解调器，其最高数据传输率为 56kb/s。

（2）ADSL 连接
目前应用最广泛的另一种接入因特网的拨号连接，是非对称用户线 ADSL。有时 ADSL 又称宽带接入，因为它比普通调制解调器所提供的带宽更宽。

ADSL 是 20 世纪 80 年代电信行业为电缆行业提供视频点播的一种解决方案，首先开发出来的。然而到 90 年代中期就认识到，作为可行的技术，应该能访问像因特网这样的高速网络。ADSL 提供非对称传输速率，典型的顺流可达到 9Mbps（从中心局到用户设备——下行），逆流在 16Kbps～640Kbps（从用户设备到中心局——上行），如图 3-5 所示。像所有的铜线传输系统一样，速率越高，作用范围越短。ADSL 传输的限制是距离，它只能工作在距电话交换局 3 英里的范围之内，而距离越是接近 3 英里，其传输速率下降得越厉害。

（3）专线连接
拨号连接只是在你需要接入因特网时才连接到你的 ISP 计算机上，而专线连接一直使你连到因特网上。使用专线连接通常要为你的 PC 安排一个静态（不变化）IP 地址，以便通过因特网来回传送数据。

专线因特网连接包括通过学校或办公室的局域网，以及 ADSL、电缆、卫星和固定无线连接等连接方式。

（4）无线连接
无线连接不使用具体的介质去连接接收和发送设备，而是使用空气本身。用于无线连接的主要技术有红外、广播无线电、微波、卫星和移动无线连接。

① 红外：红外使用红外光波进行短距离通信。由于这种光波只能直线传播，因此它有时也叫作视距通信。这就要求收、发设备能相互看清楚而中间无任何物体遮挡视线。一种最常用的技术是从笔记本计算机或个人数字助理这样的便携设备向台式计算机传输数据和信息。

② 广播无线电：广播无线电使用无线信号与无线设备通信。例如，蜂窝电话和很多可访问万维网的设备，通过广播无线电用电话呼叫与万维网连接。某些最终用户把他们的笔记本或手持电脑与移动电话连接，从远处访问万维网。这些可访问万维网的大多数

设备都采用称为 Wi-Fi（无线保真）的标准。这一无线标准广泛用于计算机相互连接，并连向因特网。

③ 微波：微波通信采用高频无线电波。与红外一样，微波由于只能直线传播，因而也是一种视距通信。因为微波只能短距离传输，所以它是城市中或大型校园内楼宇之间传输数据的很好的媒体。

蓝牙是短距离无线通信标准，使用微波进行短距离数据传输，最长距离约 33 英尺。与传统微波不同的是，蓝牙不要求视距通信。它使用的无线电波可以透过墙壁和其他非金属壁垒。

④ 卫星：卫星用于收、发大量数据。术语上行链路是指向卫星发送数据，下行链路是指从卫星接收数据。卫星通信的主要缺点是恶劣天气有时会使数据流中断。

卫星通信最有趣的一种应用是全球定位。这是一个由国防部（美）拥有和管理的，由 24 颗卫星构成的网络，它连续地向地球发送位置信息。全球定位系统（GPS）设备使用该信息去确定本设备的唯一地理位置。

⑤ 移动无线连接。

与卫星和固定无线连接使用电缆将调制解调器连向某种固定收发器不一样，移动无线连接允许设备从一处移向另一处。因此大多数手持 PC 和其他移动设备（像具有 Web 功能的蜂窝式电话）都可以使用移动无线连接并通过类似于蜂窝电话和报文设备那样的无线网络访问因特网。

第二部分　因特网应用

第 4 章　传统因特网应用

4.1　万维网概述

4.1.1　关于万维网

万维网是一种由多个因特网服务器组成的大型网络，它向运行客户应用程序（如浏览器软件）的终端，提供超文本和其他服务。

万维网允许用户从动态链接信息的全球联网服务器系列中搜索、访问和下载信息。Web 客户通常通过 Web 浏览器，向服务器传送用户需要信息的请求。服务器与客户通过传输协议，通常是超文本传输协议（HTTP）进行通信。然后，服务器用统一资源定位符（URL）访问网页。搜索引擎可用来简化访问，允许用户输入某一题目的搜索条件，从而使若干 URL 返回有关所需信息的网页。

1. 提供商

访问因特网最常用的方法是通过因特网服务提供商（ISP）。这些提供商已经连入因特网并为个人访问因特网提供通路或连接。你们的学院或大学，很可能通过局域网或电话拨号连接，免费让你访问因特网。有一些公司也提供免费的因特网访问服务。

最广泛使用的商业因特网服务提供商是全国性的提供商（例如美国在线（AOL））和无线提供商。

2. 浏览器

浏览器是访问万维网资源的程序。这种软件把你连向远端计算机、打开并传输文件、显示文本和图像，并在一种工具中提供访问因特网和万维网文档的简单界面。浏览器能使你很容易地从一个网站移动到另一个网站，对万维网进行探索或冲浪。三个著名的浏览器是 Mozilla Firefox、Netscape Communications 和微软公司的因特网浏览器（见图 4-1）。

浏览器是一种基于图形用户接口的超文本客户应用程序，用于通过万维网和因特网，对无数个远程服务器上的超文本文档和其他服务进行访问。

如图 4-2 所示，当单击超链接时，所发生的是一连串十分精彩的事件，不仅包括 Web 浏览器软件，而且包括某处的 Web 服务器，并且所包含的这些事务紧紧地依赖于 HTML 语言。

3. 个人网站

你想和全世界共享某件事情吗？你是否喜欢个人网站，但又不想学习 HTML 语言？在因特网上创建你自己的主页很容易，有很多服务能帮助你开始这项工作。

万维网上的服务网站提供制作个人网页的工具。在注册到该网站之后，你就可以使用所提供的工具制作你的网页了。一旦网页制作完成，该服务网站就作为你个人网站的宿主机使用了，而且其他网站可以从世界上任何地方免费访问你的个人网站。

4.1.2 搜索引擎

1. 概述

可以想象，当走进一个杂乱无章、堆满了书的图书馆时，要找到所需的资料几乎是不可能的。随着 WWW 的迅速发展，同样也需要对网页信息进行分门别类的管理，需要记住"是什么"和"在哪里"。在 Web 技术发展的初始阶段很难查找有用的信息。

在 WWW 的早期，美国斯坦福大学的两名研究生 Jerry Yang 和 David Filo 提出按照目录来组织超链接的方法，并发现该方法确实可行。1993 年后期，这种方法被称为"WWW Jerry Yang 向导"。这个名字不久被改为 Yahoo!，第一个搜索工具就这样诞生了。

当今有许多搜索工具能帮助我们在 Web 上迅速且方便地找到所需的信息。这些工具不断地演变，不断地优胜劣汰。与其逐一解释每个搜索引擎是如何工作的，不如只对其中的少部分进行介绍，并且给出一些网页演示，以回顾现行所有可用的搜索工具。

2. 搜索引擎是如何工作的

搜索引擎按下列顺序操作：

① 慢慢浏览万维网；

② 索引；

③ 搜索。

搜索引擎要存储很多网页信息，这些信息都是用 html 本身检索的。网页是用浏览器（有时也称为网络蜘蛛）检索的——一种能自动跟踪每次网站链接的万维网浏览器。然后分析每个网页的内容，以确定如何去索引（例如，从标题或称为元标签的专用字段中提取一些字）。有关网页的数据，存储在索引数据库中，便于以后查询时使用。查询可以按单个字进行。索引的目的是尽可能快地找到信息。有些搜索引擎，如谷歌，存储所有或部分源网页（称为高速缓存）和有关该网页的信息，而其他一些搜索引擎如 AltaVista 则存储它们能找到的每页中的每个字。由于高速缓存的页是实际索引的页，故它保存的总是实际搜索的文本，因此当当前网页的内容被更新以后并且搜索的条目不再包含在其中时，这是很有用的。

当用户往搜索引擎中输入一个查询时（通常用关键字），该引擎检查它的索引，并按其自己的标准列出最匹配的网页，通常还包含文档标题的简短摘要，有时是部分文本。

4.2 电子邮件

电子邮件是在因特网上传送电子报文的手段。从前，电子邮件只由文本报文组成。现在，电子邮件一般都包括图形、照片和很多不同类型的文件附件。世界上很多人都在相互发送电子邮件。你可以向你的家庭、同事发电子邮件。所有你需要发送和接受的电子邮件，都要使用电子邮件账号，都要访问因特网，都要使用电子邮件程序。两个使用最广的电子邮件程序是微软公司的 Outlook Express 和 Mozilla Thunderbird。

典型的电子邮件报文有三个基本部分：头部，报文和签名（图 4-3），头部最先出现，一般包括以下内容。

① 地址：发信和收信人的地址，选项是接受拷贝的人。电子邮件地址有两部分（图 4-4）。第一部分为用户名，第二部分为域名，包括顶级域名。在我们的例子中，dcaots 是 Dan 的用户名。为 Dan 提供电子邮件服务的服务器是 usc.edu。该顶级域表明，提供服务的是教育机构。

② 主题：一行字，用于描述报文的题目。主题行一般在收件人检查其邮箱时显示。

③ 附件：很多电子邮件程序允许附加如文档和表格一类的文件。如果报文有附件，该文件名出现在附件行上。

紧接着下面是信件或报文。信件一般较短而且是扼要的。最后签名行是有关发件人的信息，一般包括发件人的姓名、地址和电话号码。

在你个人或业务活动中，电子邮件是有价财产。当然，像其他很多有价技术一样，电子邮件也有一些不足之处。美国人每年都会收到几十亿个不希望和未经请求的电子邮件。这种不受欢迎的邮件称为垃圾邮件。垃圾邮件不仅是一种骚扰，而且还能造成危害。例如计算机病毒或者有害的程序，经常附着在未经请求的电子邮件上。

在控制垃圾邮件方面，反垃圾邮件已经加入到法律体系中。例如，最近制定的 CAN-SPAM Act 要求每个与市场有关的电子邮件提供一个退出选项（opt-out option）。当这一选项被选中，则接收者的电子邮件地址就从今后的邮件地址表中删除。这种方法使用以来，从美国境外服务器来的垃圾邮件已经减少了 50%。另一种更为有效的方法已经开发了，并且已经使用了，就是垃圾邮件拦截器（见图 4-5），这些程序使用各种不同的方法去识别和删除垃圾邮件。

4.3 电子商务和物联网

4.3.1 电子商务

1. 什么是电子商务

电子商务是一个系统，它不仅包括可以直接产生收益的那些商品交易和服务，而且也包括那些支持产生收益的事务处理，例如为那些商品和服务所产生的需求，提供销售支持和客户服务（见图 4-6）或提供商业伙伴之间的通信等。

电子商务是在传统商业的优势和结构上，增加了计算机网络提供的种种灵活性而建立的。

电子商务产生了一些新的商业模式和一些新的运营方式。例如，亚马逊 Amazon.com 是一家在华盛顿州西雅图的图书经销商。这家公司没有实际的书店，而是通过因特网售书，由与其合作的出版商直接发送图书，因而，该公司不必拥有任何库存。而像 Kantara 和 software.net 这样的公司在此方面更进了一步。

因为他们的所有产品（商业软件包）都是电子的，都能存储在相同的计算机中，这些计算机也用于处理订单，并作为 Web 服务器使用，其库存则全部是数字化的。作为另外一个例子，AMP 公司正向其客户提供这样的机会：客户可以直接从它的 Web 服务器目录上选购电子接插件和相关配件，而无须基于电子数据交换 EDI 的订购和确认。

2. 电子商务业务模式

一个公司的政策、运作和技术决定了它的业务模式。本质上，一个公司的业务模式描述了该公司如何产生收益。有很多标准电子商务业务模式，包括企业对消费者（B2C）、企业对企业（B2B）、消费者对消费者（C2C）和企业对政府（B2G）等模式。下面讨论这些最常用的模式。

（1）企业对消费者（B2C）

企业采用 B2C 模式向单个消费者销售物品或提供服务。B2C 模式是最早定义并通过万维网实现的主要电子商务业务模式之一。B2C 企业的几个例子包括亚马逊（Amazon.com）、宾恩（L.L.Bean）、沃尔玛（Walmart.com）和 Polo.com（见图 4-7）。这些企业可能只是因特网商店，也可能是既在网上又有店铺的混合经营商店。

（2）企业对企业（B2B）

企业对企业的应用包括发生在两个企业之间的任何类型的电子商务交易。B2B 的收益一直在增长，并有望在今后几年内持续大幅度增长。

（3）消费者对消费者（C2C）

消费者对消费者——有时也称为个人对个人（P2P）——其业务模式大多数只是由消费者拍卖而形成，即这里的消费者向其他消费者销售产品。易趣（eBay）购物网是当今最大的 C2C 电子商务企业之一，每天有上百万件产品被拍卖。

（4）企业对政府（B2G）

美国政府每年开支超过 5000 亿美元，这种趋势并未见减弱，B2G 组织正在变得异常重要。这些组织向本地、州和联邦政府购买者销售产品和提供服务。一般来讲，政府部门接受在线购物的程度比私人要差。

与 B2G 有关的一些措施包括美国有些州允许市民网上付款，如付税、更新驾照等。这有时称为客户对政府（C2G）电子商务。

4.3.2 物联网

1. 物联网概述

物联网是具体的物品或"东西"的网，这些物品嵌入了电子器件、软件、传感器和连接，通过与制造厂商、操作人员或其他连接设备交换数据，获得更大的价值和服务。每件物品可通过嵌入式计算机系统识别，而且在现在的因特网设施内都是可以互操作的。

典型的物联网有望提供先进的设备、系统和服务的连通性，并超越机器对机器间的通信，覆盖各种协议，覆盖各个域和应用。这些嵌入设备（包括智能物品）的互联，有望在自动化领域的各个方面起引领作用，同时也使像智能网格这样的技术有实现的可能。

物联网中的东西，可以是各式各样的设备，如心脏监护移植设备、安在农场动物身上的生物芯片收发报机、海水中的电子蚌、带有内置传感器的汽车、帮助消防队员搜救的现场操作设备。这些设备在现有的各种技术的协助下，收集有用的数据，这些数据以后自动在其他设备之间自动流动。当前市场上的产品有智能调温系统，以及利用 Wi-Fi 的远程监视的洗衣机和烘干机。

除了因特网与自动化系统的连接已经扩展到许多新的应用领域之外，物联网有望从各个地方产生大量的数据，并且数据聚集的速度非常之快，因而需要制作索引，要存储和处理这些数据。

2. 物联网的体系结构

物联网系统理应有一个事件驱动的体系结构。图 4-8 是物联网发展的三层体系结构。顶层是由驱动应用程序组成。这个应用空间非常大。底层代表不同类型的传感器：主要是射频识别标签、ZigBee 或其他类型的传感器，以及标识路线图的 GPS 导航仪。收集在这些传感器上的信号或信息，通过中间层上的云计算平台与应用程序相链接。

在中间层的移动网络上，在因特网主干网上和各种信息网上，构建了信号处理云。在物联网中，感知事件的含义是不遵循确定性或句法模型的。事实上，这里采用的是面向服务的体系结构模型。大量传感器和滤波器用来收集原始数据。各种计算和存储云以及网格用于处理数据，并将其转换为信息和知识格式。所感知的信息可共同为智能应用构建一个决策支撑系统。也可以把中间层视为语义万维网。有些行为体（服务、组件、计算机化身）是自引用的。

3. 应用

按照 Gartner 公司（技术研究和顾问公司）的说法，到 2020 年物联网上的设备将接近 260 亿台，ABI 的研究估计，到 2020 年将超过 300 亿台设备是用无线连入物联网的（万物互联）。Pew Research Internet 课题最近的观察和研究表明，绝大多数技术专家和有回应的因特网用户（约占 83%）同意这种观点，即到 2025 年，物联网和物联云，嵌入式和可穿戴计算设备（以及相应的动态系统）将有广泛的、重大的影响。因此，很清楚，物联网将由非常大量的连入因特网的设备组成。

具有有限 CPU、存储器和电源的网络嵌入设备的能力表明，物联网可在任何领域找到应用。这些系统应该负责收集所设置的信息，范围从自然生态系统到建筑物和工厂，并由此，在环境监测和城市规划领域中找到应用。另一方面，物联网系统也应负责完成不仅是感知物体的任务，例如，智能购物系统，还应该通过跟踪用户手机，监视某些用户在商店的购物习惯。然后专门向用户提供他们喜欢的产品，甚至报告给他们所需物品在什么地方。另外一些感知和执行的应用例子是与热、电和能量管理有关，还有旅游辅助交通系统等。

当然，物联网应用不仅限于这些领域，也还有其他物联网专用场合。在这里把其他一些最突出的应用领域概述一下。基于应用，物联网产品总体可分为五类：智能穿戴、智能家庭、智能城市、智能环境和智能企业。在这些市场中，每一类物联网产品和解决方案都有各自的特性。

第 5 章　因特网新应用

5.1　即时消息

5.1.1　QQ

1．QQ 概述

腾讯 QQ 通常称为 QQ，是中国大陆上最普通的免费即时消息计算机程序。到 2010 年 9 月 30 日，QQ 即时消息实际用户数大约 6.37 亿，可能是世界上最大的在线团体。同时在线的 QQ 用户数超过 1 亿。2011 年 2 月，QQ.com 在 Alexa 因特网排行榜中名列第 10，仅排在第 9 位的 Twitter 之后。该程序由 Tencent Holding Limited（HKEX 0700）维护，Naspers 有部分版权。自从 QQ 进入中国家庭以来，很快就成为现代的一种文化现象，现在正成为一种大众文化。除了这个聊天程序之外，QQ 还开发了许多包括游戏、虚拟宠物、铃声下载、音乐、购物、博客、微博、成组语音聊天等在内的子功能。

QQ 当前版本是 QQ 2010 beta 2。腾讯周期性地公布 QQ 专用版，以配合如奥林匹克运动会或中国新年这些事件。

正式客户运行在微软的 Windows 上，而 beta 大众版是为 Mac OSX 10.4.9 版或更新版开发的。万维网版，WebQQ（整版）和 WebQQ Mini（小型版），是用 Ajax 编写的，这些 QQ 现在都可以使用。

截至 2015 年 1 月，腾讯已拥有 8.29 亿账户，最高峰同时在线 QQ 用户为 1.76 亿。

2．QQ 国际

（1）Windows

从 2009 年开始，QQ 通过专用英文门户网站，将其 Windows QQ 国际客户端扩展到国际服务。

QQ 国际是为非本土语言交谈者提供使用与中国本土使用者一样的所有功能的一种环境。通过交谈和视频呼叫与其他 QQ 用户接触，并向非本土用户提供访问腾讯社交网（Qzone）的界面。该客户端支持英语、法语、西班牙语、德语、韩语、日语和普通汉语。第三方应用软件这一财富与 QQ 国际绑在一起有助于更方便地与国内外用户跨文化交流。

QQ 国际的一个主要功能是有一个选项，它可以自动地对所有交谈进行机器翻译。

（2）Android

QQ 国际的安卓版是 2013 年 9 月公布的。该客户端界面可使用英语、法语、西班牙语、德语、韩文、日文和普通汉语。除文本消息外，用户间可以相互传送图像、视频和语音媒体消息。另外，用户通过此客户端的 Qzone 界面与所有接触的人共享多媒体内容。

现场翻译功能对所有进来的消息都可用，并支持 18 种语言。

（3）iOS/ iPhone

2013 年底，用于 iPhone 和 iOS 设备的 QQ 国际公布了，与安卓所宣布的完全一致。

3．Web QQ

2009 年 9 月 15 日，腾讯正式宣布它的基于 Web 的 QQ，最新版本是 3.0。与很少用的基于 Web 的即时消息不同，Web QQ 3.0 的功能更像它自己的操作系统，并在桌面上可以加入 Web 应用软件。

4．开源与跨平台客户端

使用逆向工程技术，从事开源任务的人们能够很容易地了解 QQ 协议，并着力去实现与更多的用户友好的客户端相兼容的客户核心库的功能。这些客户端大多数是跨平台的，因此它们能够使用官方客户端不支持的那些操作系统。然而，这些功能只是官方客户软件功能的子集，因而功能有限。

5.1.2 脸谱

1．特性

脸谱是一种社交网络服务，也是一个网站，是 2004 年 2 月公布和运作的，属于 Facebook 公司。到 2014 年 6 月，脸谱的实际用户已超过 13 亿。用户可以创建带有照片、个人兴趣目录、联系信息和其他个人信息的个人配置文件，还可将其他用户作为朋友加进来，当他们更新他们的配置文件时，可以交换信息，包括自动公告。另外，用户可以加入共同感兴趣的用户组，这些是由车间、学校或院校组织的，也可以按其他特点去组织。

脸谱有很多用户可以进行交互的特性：

（1）Wall，是每个用户配置文件页上的一个空间，允许他的朋友向该用户发送信息，供查看。

（2）Pokes，允许用户相互发送虚拟"戳"（一个公告，用于告诉用户他们已经存入了"戳"）。

（3）照片，用户可以上传照相册和照片。

（4）状态，用户可以告之他们的朋友，他在哪里，在干什么。

根据隐私设置，可以查看用户配置文件的任何人，都可以看那个用户的 Wall。

2010年2月23日，脸谱在它的News Feed的某些方面被授予专利，该专利覆盖可以提供链接的News Feed，因此一个用户可以参加另一个用户的同一活动。这一专利可以促使脸谱对违反专利的那些网站采取行动，这些网站很可能包括像推特那样的网站。

脸谱最常用的应用之一是照片应用——用户可以上传相册和照片，脸谱允许用户上传无限量的照片，而其他图像托管服务，如Photobucket和Flickr则限制了用户上传照片的数量。在前几年，脸谱限制用户每本相册60幅照片。到2009年5月，每本相册可有200幅照片。

2．技术

脸谱是用PHP构建的，这里PHP是用HipHop for PHP这种"源代码翻译程序"编译的，而这个源代码翻译程序又是由脸谱工程师编写的，他们又将PHP转换成C++。据报告，使用了HipHop之后，脸谱服务器的CPU占用时间减少了50%。

脸谱使用了基于HBase的组合平台，在分布式机器上存储数据。由于采用拖尾架构，新事件可以以日志文件形式存储，而且日志是拖尾的。该系统弹出这些事件并将他们写入存储器。此后，用户界面将这些数据提出并将其显示给用户。脸谱习惯处理AJAX请求，这些请求用Scribe（由脸谱开发的）写入日志文件。

5.1.3 推特

推特是一种社交网在线服务，它能使用户发送和读取称为"推文"的140个字符的短消息。

注册了的用户可以读取和发送推文，但非注册用户只能读取消息。用户可以通过推特网站界面、短信服务（SMS）或移动设备App访问推特。推特公司位于美国旧金山，在全世界有超过25个办公室。

推特创建于2006年3月，其业务迅速普及全世界，2012年用户超过1亿，每天发送推文3.4亿个，每天处理16亿个搜索请求。2013年，推特成为10个访问最多的网站之一，并作为"因特网的短信服务"来表述。到2014年12月，推特用户已超过5亿，其中超过2.84亿是活跃用户。

1．特性

推特在默认情况下是公开可见的，但发送者可以限制消息发送给某些人。用户可以通过推特网站，与外部兼容的应用（如智能电话），或通过某些国家的短信服务使用推文。用户通过推特向前转发推文的过程叫作转推。无论是推文还是转推文，都可以被跟踪，以观察哪些用户是用得最多的。由于这种服务是免费的，通过短信服务访问它，可能会产生电话业务提供商的费用。

推特允许用户用手机去更新他们的配置文件，可以通过智能电话和台式机中的文本消息，或应用程序去更新。

作为一个社交网络，推特围绕着跟进者这个概念在发展。当你打算去跟进另一个推特用户时，那个用户所发布的推文就会按照逆向顺序出现在你的推特主页上。假设你跟进了20个人，那么你可能在滚动页面上看到这20个用户的混合在一起的推文，比如会有有关谷物早餐的消息、感兴趣的网址连接、音乐推荐，甚至对未来教育的假想。

2．实现

推特在很大程度上依赖于开源软件。推特 Web 界面采用 Ruby on Rails 框架,安排在 Ruby 的增强型企业版功能上。

到 2011 年 4 月 6 日,推特工程师们决定将他们的 Ruby on Rails 搜索栈转换到他们称为 Blender 的 Java 服务器上。

该服务的应用程序界面允许其他 Web 服务和应用与推特集成在一起。

5.1.4 微信

微信是由中国腾讯公司开发的一种移动文本和话音消息通信业务,2011 年 1 月首次推出。按每月的实际用户来看,它是最大的独立消息应用软件。

这项应用可以在安卓、iPhone、黑莓、Windows Phone 和 Symbian 电话中使用,并可以在基于 Web 和 OS X 的客户端使用,但这些都要求用户在对该应用有授权的手机上使用。截至 2014 年 8 月,微信用户已有 4.38 亿,其中 0.7 亿在境外。

用户注册微信可以用脸谱账号或电话号码。目前用电话号码注册的国家超过 100 个。不能通过腾讯 QQ 直接注册。但通过电话号码注册之后,用户就可以用腾讯 QQ 账号连接他们微信的账号了。

使用微信,可以传送文本消息,进行"按住说"通话,广播(一对多)消息,共享照片和视频以及位置共享。微信通过蓝牙技术让人们相互联系,并在需要时随机提供联系的各种手段,还可以和社交网络服务集成,比如通过脸谱和腾讯 QQ 运行的那些微信。照片还可以用滤波软件和文字说明去修饰,也可以使用机器翻译服务。

微信支持用户使用公共账号去注册,这使得用户将馈送送给其他用户,与他们互动,为他们提供服务。到 2014 年底,微信公共账户已达到 800 万。

在中国,微信公共账户已成为政府、新闻媒体和公司的普通业务,也是推广的平台。专用公共账户的用户使用该平台可进行医院预约、签证更新或信用卡等业务。

2014 年 9 月 30 日,公布了带有视力捕获和共享等新功能的 WeChat 6.0。

按 GlobalWebIndex 统计,微信是世界上第五个用得最多的智能电话应用软件,2013 年 8 月,它紧随 Google Maps、Facebook、YouTube 和 Google+之后。微信还宣布,它有 1 亿个国际注册用户,这是仅用了 3 个月,就从 5000 万上升到 1 亿的,而中国用户为 3 亿。

按新华社的统计,2013 年 10 月,全世界微信用户为 6 亿,而且全部用户的 30%是在国外。

5.2 社交网络服务

5.2.1 维基

维基是一个通过万维网浏览器,使用简化的标记语言,或所见即所得(WYSIWYG)文本编辑器,创建和编辑任何数量的相互链接网页的一个网站。维基通常由 wiki 软件授权,并且经常由多个用户联合使用。应用维基的例子有团体网站、公司内联网、知识管理系统和注记服务，该软件也可用于个人笔记。

维基可用于不同的目的。有些可以控制不同的功能（访问级别），例如，编辑权可对材料进行更改、增加或删除。其他功能则不要求强制访问控制就能进行访问。

维基概念的本质是：

（1）维基吸引所有用户在维基网站内去编辑网页或创建新的网页，它只使用 plain-vanilla 万维网浏览器，而不需要任何额外的附件。

（2）维基鼓励在不同网页之间的有意义的题目联合，方法是使网页链接操作很直观且容易，并显示是否有预想的目标网页存在。

（3）维基对偶尔的访客不是一个精心设计的网站。相反，它试图把访客引入到一种创作和合作的过程中，并经常改变万维网网站的景色。

在 wiki 网站中的单一网页称为"维基页"，而全部网页的集合称为"维基"，这些网页可以通过超链接很好地互联在一起。维基本质上是用于创建、浏览和搜索信息的数据库。

维基技术很明显的一个特性是使用它制作和更新网页很容易。一般来讲，在修改被承认之前无须检查，很多维基是公开允许公众去交流的，不要求他们去注册用户账户。有时为了会话则建议去注册，以便建立一个"维基—签名"网络跟踪器，用于签署自动编辑。但是，很多编辑可以实时构建并能即时在线出现，这可能助长该系统的违规使用。私人维基服务器要求对编辑网页进行用户验证，有时甚至读网页也要用户确认。

维基也是一种应用程序，是一个典型的万维网应用软件。它允许合作式修改、扩充或删除它的内容和结构。在典型的维基中，书写文本可以使用简单的标记语言（称为"维基标记"）或功能强的文本编辑程序。虽然维基是一种内容管理系统，但它还是与博客或其他类似的系统不同，区别在于所书写的内容无须规定拥有者或领导者，而且维基有一点隐含结构，允许在用户需要时才显现出来。

Wikipedia 这一百科项目，是能在公共万维网上浏览网页的最普通维基，但是有很多网站是运行多种不同的维基软件的。维基可以用于多种网站，既可是公众的，也可是私人的，包括知识管理、做笔记、团体网站和因特内联网。有些维基还可以对不同的功能（访问级别）进行控制，例如，某些编辑权限允许对材料进行变更、增加或删除。有一些则允许在没有强制访问控制下进行访问。而另外一些规则会迫使你对内容进行组织。

5.2.2 博客与微博

1. 博客

博客（Web 与 log 的混成词）是一种万维网站或网站的一部分。博客一般由个人维护，并有评论、事件描述的词条或像图形或视频那样的其他材料。词条一般以反时间顺序显示。博客也可以作动词使用，意为维护博客或增加内容到博客上。

大多数博客是交互式的，可以通过博客上的小窗口相互留言，甚至报文。这种交互与其他静态网站的交互是有区别的。

很多博客是按具体主题提供评论或新闻的，而其他博客则负责更个人化的在线日记工作。典型的博客可以把文本和图像组合在一起并链接到其他博客、网页和其他与本题目相关的媒体上。读博客的人可以把评论以交互的格式留下来，这是很多博客的重要功能。虽然某些博客集中在艺术（艺博）、摄影（影博）、视频（视博）、音乐（MP3 博）和语音上（podcasting），但大多数博客主要是文本形式的。微博则是博客的另一种类型，是一种很短的邮件。

截至2011年2月16日，公共博客已超过1.56亿。到2014年2月20日，世界上大约有1.72亿 Tumblr 博客和7580万 WordPress 博客。按照某些评论家和博主的说法，博主是当今最普通的博客服务使用者，但是博主不提供公共统计数字。到2014年2月22日，Technorati 有130万博客。

有很多不同类型的博客，他们之间的区别不仅在内容的类型上，分发和书写内容的方法上也有所不同。

2．微博

微博是一种博客形式的广播媒体。微博与传统博客的不同之处是它的文件规模比较小。微博"允许用户交换像短句、单个图像或视频链接那样内容的小元素"。

与传统博客相比，微博版主发出的题目范围，从简单的，如"我现在正在做什么"到"赛车"这样的主题。也有商业微博，用于促进网站服务或促销产品，并促进一个单位内的合作。

有些微博服务提供隐私设定这样的功能，它允许用户去控制谁可以读他们的微博；另外，除了基于 Web 的界面，还可用另外一些方法公布一些事项。这些可能包括文本消息、即时消息、电子邮件或数字声音。

微博服务已经改变了信息消费的方法。它已经使市民本身强化为数据的感知者或数据源，而数据也可能变成重要的信息片段。人们现在共享在他们周围观察到的东西，共享有关事件的信息，共享某些方面的见解，例如，共享政府的健康管理政策。

还有，这些服务存储了从这些邮件来的各种各样的元数据，如位置和时间。这种数据的汇聚分析包括各个方面，像空间、时间、题目、情趣、网络结构等，并给研究人员一个机会，以便从某些感兴趣的事件中了解人对社会的认知。微博也刺激了著述业，在 Tumblr 微博平台上，再博客功能将邮件链接到原创那里。

微博有可能成为一种新的非正式的通信媒体，特别适合于单位内合作工作的情况。过去几年，通信方式已从面对面转移到网络电子邮件、即时消息、文本消息和其他一些方式上。但是，有人认为，现在的电子邮件是一种慢速、低效率的通信方法。例如，耗时的"电子邮件链"可以发展下去，但是，两个或多个人可能会为简单的事情，如安排会议，而消耗在冗长的通信中。微博提供的一对多广播方式，可以认为由于回避了这种不利情况而提高了效率。

用户和单位可以建立他们自己的微博服务：为此可以使用免费和开源软件。托管微博平台也可用于商业和机关。

5.3 云计算

云计算是一种交付服务而不是产品，它提供了一种计算方式，在这种方式中，将共享的资源、软件和信息作为一种实用程序通过网络，特别是因特网提供给计算机或其设备（这类似于电网）。

1. 概述

云计算在技术上是一种商业术语，这种技术提供计算服务、软件、数据访问和存储服务，而不要求最终用户知道提供这些服务的系统所在的实际位置和配置。与这一概念相似的是电网，电网内的最终用户用电，也不必知道提供服务的供电设备或基础设施的情况。

云计算是一种基于因特网协议的 IT 业务的新的补充、消费和交付的新模型，通常包含动态可伸缩的，并且是虚拟化的资源。它是一种由因特网提供的，远程计算站容易访问的副产品和后续产品。这可以采取基于 Web 工具的形式，或者是通过 Web 浏览器，用户能访问和使用的应用程序，如果该应用程序已经安装在用户本地的计算机上了。

云计算提供商通过因特网交付应用，这些应用程序可以从 Web 浏览器、台式计算机和移动应用上访问，同时，商业软件和数据存储在远端的服务器上。某些情况下，继承应用（是一种商业应用系列，直到现在，在瘦客户 Windows 计算中，还是很普遍的）是通过共享屏幕技术提供的，而计算资源是统一放在远程数据中心上的；其他情况，整个商业应用是用像 AJAX 那样的基于 Web 的技术编码的。

基础会聚（会聚的基础）和共享服务的更广义的概念是云计算的基础。这种类型的数据中心环境能使企业去建立他们自己的应用，并更快地运行起来，而管理比较容易，不需要太多的维护，也能使 IT 业更快地调整 IT 资源（如服务器、存储器和联网），以满足变动的、不可预期的商业需求。

大多数云计算的基础设施是由服务组成的，这些服务是通过共享数据中心交付的，并对消费者的计算需求是以单一访问点呈现的。

云计算对商业的巨大影响已经促使美国联邦政府把云（计算）视为重构他们的 IT 基础和降低预算的措施。随着顶级政府正式采用云计算，很多代理商已经至少有一个或多个云系统在线了。

图 5-1 为云计算逻辑图。

2. 公共云、私有云和混合云

公共云是构建在因特网上的，凡是付费给这种服务的用户都可以访问它。公共云归服务提供商所有，通过预约就可以使用。图 5-2 顶部的标注框是典型的公共云架构。可以使用的公共云有很多，包括谷歌应用引擎（GAE）、亚马逊万维网服务（AWS）、微软的 Azure、IBM 的蓝云和 Salesforce.com 的 Force.com。上述云的提供者是商业性的，他们为在他们的专利架构内创建和管理虚拟机，提供一个公共的远程访问接口。公共云提供一套可选择的商业流程。这种应用和架构服务是基于柔性价格的。

私有云是构建在一个单位所拥有的内联网范围内的。这样，它就是客户拥有和管理的，对其访问也就受限于它所拥有的客户和伙伴。这种安排并不意味着可以通过公共访问接口在因特网上出售容量。私有云为本地用户提供一个柔性的、灵活的私有架构，在它管辖的范围内承载工作负荷。私有云可以视为一种更有效、更舒适的云服务。它可能影响着云的标准化，但它又保持着更多的用户和单位的管理权。基于内联网的私有云连入公共云是为了获取更多的资源。

混合云是公共云和私有云二者共同构建的云，如图 5-2 左下角所示。私有云也能支持混合云模型，其支持方式是使外部公共云的计算能力为本地服务。例如，研究计算云（RC2）是由 IBM 公司构建的一种私有云，它连接散布在美国、欧洲和亚洲的八个 IBM 研究中心的计算和信息技术资源。混合云可对客户、合伙人网和第三方进行访问。

总之，公共云激励着标准化，维护资本投资，提供应用的灵活性。私有云则力图做到用户化，为用户提供高效、弹性、安全和保护私密的环境。混合云处于这二者之间，借助于资源共享而采取了很多折中。

3．特性

云计算展示出以下重要特性。

（1）计算资源授权给最终用户，方法是将这些资源放在用户自己的控制之下，而不是由中央式的 IT 服务去控制。

（2）灵活性改善了用户反复使用基础技术资源的能力。

（3）应用程序接口对软件的访问能力，能使机器与云软件交互，其方法与用户接口能使人与机器交互的方法一样。云计算系统一般使用基于 REST 的 API。

（4）设备和位置独立性能使用户用 Web 浏览器访问系统，而不管系统的位置或使用的是何种设备（PC 还是手机）。当基础设施不在网站上，（典型的是由第三方提供的设备），并且要通过因特网访问时，用户可以从任何地方进行连接。

（5）多租用性，允许跨越大的用户群共享资源和分摊费用。

（6）可靠性增加了，如果使用多个冗余网站的话，这种良好设计的云计算很适合于企业运作的连续性和灾害恢复。

（7）性能是可监视的，使用 Web 服务这种系统接口，可构建连续而松散的耦合体系结构。

（8）安全得到改善，这是由于采取了数据集中管理，强化资源的安全等措施。

（9）云计算应用的维护比较容易，因为这些应用软件不需要安装在每个用户的计算机上。

5.4 大数据

大数据是一种表示数据集的、意义广泛的术语。这种数据之大、之复杂，是传统数据处理应用所无能为力的。这种数据的挑战性包括分析、截取、处置、搜索、共享、存储、传送、可视化和信息隐匿。简言之，这一术语是指使用预测分析或其他一些先进的方法，从数据中提取其值，但又很少关注数据集的具体规模。

通过对数据集的分析可以发现"现货企业的趋势，灾害的预防，暴力犯罪等"问题中的新的相关性。在很多领域中，科学家、媒体和广告从业者以及政府都遭遇到大数据集的限制。这些限制影响了因特网的搜索，财务和企业的信息工作。

例如，科学家在电子科学领域，包括气象学、基因组学、神经网络体学、复杂物理模拟和生物及环境科学研究中就受到限制。

数据集在规模上的增长，部分原因是数据持续从各种来源被广泛收集，这些来源包括廉价而大量的移动设备、高空传感技术（遥感）、软件记录、相机、麦克风、射频标识阅读器和无线传感器网。世界技术的发展，人均存储的信息量，自 20 世纪 80 年代以来，大概每 40 个月翻一番。到 2012 年，每天产生 2.5 艾字节个数据。对大型企业的挑战是确定谁将拥有跨越整个机构的大数据的主导权。

图 5-3 为整体信息存储量的增长情况。

1．定义

大数据是用来描述信息（结构化和非结构化的）的指数增长，有效性和使用的通用术语。按国际数据集团的说法，是单位和信息行业的领导们特别关注的，不断增加的信息量、信息种类和传播速度，促成了大数据的形成。

（1）量

很多因素造成了数据量的增加——长年存储的业务数据，经常性的、从社交媒体来的文本数据流，收集的传感器数据等。以前，过多的数据量造成存储问题。虽然现今的存储价格降低了，而其他问题又出来了。例如，如何确定大量数据之间的关联性问题，如何从有关的数据中产生新值的问题。

（2）种类

当今，数据以各种形式出现——从传统数据库到由最终用户和联机分析处理软件产生的层次型数据存储，到文本文档、电子邮件、计量器收集的数据、视频、音频、股票行情指示器和财经业务数据。据估计，约 80%的单位，数据不是数值化的！但这些数据必须包含在分析和决策支撑之中。

（3）速度

按照甘特的说法，速度是指"产生数据的速度有多快，满足要求的数据处理速度有多快。"射频识别标记和智能仪表是使数据接近于实时迸发的推手。

2．大数据的使用

实际问题不是你收集了大量数据（因为很清楚，我们已经处于大数据时代了），而是用那些关系重大的大数据去做什么。大数据的有用之处是，单位有能力去利用相关联的数据，并用这些数据去做出最好的决策。

当今的技术不仅支持大量数据的收集和存储，也提供了解和利用其全部值的功能，这有助于单位的工作更加有效并获益。例如，利用大数据和大数据解析学，可以：

- 分析数据量庞大的库存，以确定最大赢利下的最优价格并清仓。
- 几分钟内重新计算有价证券的风险，了解今后降低风险的可能性。
- 快速确认关系重大的那些客户。
- 在销售点，根据客户当前和以前的购物，产生零售优待券，确保较高的偿还率。
- 在恰当的时间，向移动设备发出合适的建议，而客户能在合适的地点利用这些建议。
- 分析从社交媒体来的数据，以便在需求方面探测新的市场趋势。
- 使用单击流分析和数据挖掘技术，检测欺骗行为。
- 通过研究用户行为、网络运行记录和机器传感器，确定产生事故、问题和缺陷的根源。

大数据的一些实例：
- 射频识别系统产生的数据量是传统条形码系统的 1 千倍。
- 全世界每秒钟进行 1 万笔支付卡交易。
- 沃尔玛每小时处理 1 百多万笔客户交易。
- 每天传送的推文多达 3.4 亿个，相当于每秒 4 千个推文。
- 脸谱有超过 9.01 亿个产生社交数据的实际用户。
- 超过 50 亿人正在手机上呼叫，传送电文，发推文和浏览网站。

3. 技术

大量新技术的使用，使得很多单位产生了大量大数据，并且需要对这些大数据进行分析，这是因为：
- 廉价的海量存储器和有强大处理能力的服务器。
- 更快的处理器。
- 能买得起的大容量存储器，如 Hadoop。
- 专门为大数据量（包括非结构化数据）设计的新存储和处理技术。
- 并行处理，簇技术，大规模并行处理，虚拟化，大栅格环境，高连通性和高吞吐率。
- 云计算和其他灵活的资源分配和配置。

大数据技术不仅具有收集大量数据的能力，也具有了解大数据和利用它的值的能力。所有单位利用大数据集的目标，应该是利用最相关的数据，进行最优化的决策。

非常重要的一点是，并非所有的数据都是相关或有用的。但如何发现关系最密切的数据点？这是一个大家都认可的问题。"大多数企业已经经历了从大数据中提取有效值的漫长过程。有些公司企图在大数据上采用传统的数据管理方法，只学会了不再使用陈旧的规则"，这是 Dan Briody 2011 年在经济学家智库刊物上说的，"大数据利用游戏变更财富"。

第三部分　程序设计语言和数据库

第 6 章　程序设计语言

6.1　C、C++和 C#

C 语言把结构化高级语言的一些最好的特性与汇编语言组合在一起了——也就是它编程相对容易（至少与汇编语言相比较），能有效地使用计算机资源。虽然 C 最初是作为系统程序设计语言设计的(实际上 UNIX 操作系统的主要程序首先是用 C 语言写的)，但它已被证明是可用于各种应用软件的强大的和灵活的语言。C 语言是绝大多数计算机专业人员用来开发软件产品的语言。

比较新的、面向对象的 C 版本被称为 C++（见图 6-1）。C++包括 C 的基本特性，所有 C 程序能被 C++编译器识别。但 C++具有另外一些特性，如对象、类和面向对象程序的其他元素。C++也有可视化版本（VC++）。所有这些表明，C++是用于图形应用的最通用的程序设计语言之一。

C++完全支持面向对象的程序设计，包括面向对象开发的四个特征：封装、数据隐藏、继承性和多态性。

单元的自包含特性称为封装。使用封装，我们就可以实现数据隐藏。数据隐藏是一个很有价值的特征，这样，用户可以使用一个对象而无须知道或关心其内部的工作机制。这就像你使用冰箱，而不需要知道压缩机是如何工作的一样，你可以使用一个设计得很好的对象，而不用知道它的内部数据成员的情况。

通过创建用户自定义的类，C++支持封装和数据隐藏特性。一旦一个良好定义的类，作为完全封装的实体被创建，就可以充当一个整体来使用。这个类的内部真实工作被隐藏起来了。使用良好定义的类的用户，不需要知道类是如何工作的，他们只需要知道如何使用它就行。当 Acme 汽车公司的工程师们想制造新车的时候，他们有两种选择：要么从零开始，要么对现有的车型加以修改。也许他们的星型模型是近乎完美的，但是他们想加一个涡轮充电器和一个 6 速变速器。总工程师则不想从头做起，他说："我们再造一个'星'吧，并且要增加一些功能，我们把它叫作'类星'吧。""类星"车是"星"型车的一种类型，但"类星"具有一些新的特点。

C++通过继承支持重用性。可以声明一个新类，该类是一个现有类的扩展。这个新子类可以说是从现有类派生来的，故有时称为派生类。这就好像"类星"车源自"星"车，它继承了"星"车的所有性能，但能按需增加一些新的性能。

C++支持这样一种思想，即不同的对象通过所谓的函数多态性（polymorph）和类的多态性都能做"正确的事"。poly 意思是许多，morph 意思是形状。多态性是指同样的名字有多种形式。

尽管 C++是 C 语言的超集，而且实际上任何一个合法的 C 语言程序也是一个合法的 C++程序，但是 C++与 C 之间的区别是很明显的。C++凭借与 C 语言的关系而受益多年，是因为 C 语言程序员可以很容易地掌握 C++。然而，为了真正掌握 C++特点，许多程序员发现必须抛弃以前所知道的一些知识，而学习全新的概念和解决编程问题的方法。

C++的主要标准版本之一是 C++11（正式称为 C++OX），是 2011 年 8 月 12 日获批并公布的。

2014 年 C++14（也称为 C++1y）作为 C++11 小规模扩展版本面世，其主要特点是错误的修正以及一些小的改进。其目标类似于 C++03 对 C++98 做的一样。2014 年 8 月中旬，C++14 国际标准草案投票过程完成。

在 C++14 之后，主要版本，非正式名称是 C++17，计划 2017 年推出。

C 的最新版本是 C#（读作"C 夏普"）。作为 C 和 C++的混合产品，C#是为直接与 Sun 公司的 Java 语言竞争而开发的最新的程序设计语言。C#是为提高开发 Web 应用软件效率而设计的一种面向对象的程序设计语言。C#5.0 是最新版本，是 2012 年 8 月 15 日发布的。

微软的 Visual C#也遵循 C#规范，它包含在微软 Visual Studio 产品系列中。它是基于 C#语言的 ECMA/ISO 规范的，也是微软公司的产品。虽然该规范有多个品种，但 Visual C#显然是迄今为止应用最广的。

6.2 Java

1．概述

Java 是一种程序设计语言，最初是由 James Gosling 在 Sun Microsystems 公司（现为 Oracle 公司的一部分）开发的，并于 1995 年作为 Sun Microsystems 的 Java 平台的核心软件公布的。该语言很多是来自 C 和 C++语法，但有一个比较简单的对象模型和少量的低层功能。Java 应用程序一般被编译成字节代码（类文件），字节代码可运行在任何 Java 虚拟机（JVM）上，与计算机的体系结构无关。Java 是一种通用的、并发式的、基于类的、面向对象的语言，它是专门设计成具有尽可能少的实现的依赖性的。其目标是使应用程序开发者能"写一次，便可在任何机器上运行"，这表明编译好的 Java 程序可以运行在所有支持 Java 的平台上而无须再次编译。Java 是目前最通用的程序设计语言之一，特别适合于客户-服务器 Web 应用环境，据报告有 900 万开发者。

2．目标

研制 Java 语言有 5 个主要目标：

（1）它应该是"简单、面向对象和通俗的"；
（2）它应该是"健壮和安全的"；
（3）它应该是"体系结构是中立的和可移植的"；
（4）执行时应该是"高性能的"；
（5）它应该是"解释性的，线程的和动态的"。

3．Java 平台

可移植性是 Java 的一种特性，也就是用 Java 语言编写的计算机程序可运行在任何类似的硬件/操作系统平台上。这是通过把 Java 语言代码编译成称为 Java 字节代码的中间表达式实现的，而不是直接编译成由平台规定的机器代码。Java 字节代码指令类似于机器代码，但是能被专为该宿主机硬件写的虚拟机（VM）解释。最终用户通常使用在他们自己机器上的为单独 Java 应用而安装的 Java 运行环境（JRE），或使用用于 Java applets 的 Web 浏览器。

标准化库提供访问宿主机的专用功能（如图形、线程和网络技术）的通用方法。

采用字节代码的主要好处是移植性。但是，解释所需要的开销表明，在本地可执行环境上，运行解释程序总比编译程序要慢。早期就开发出来的即时编译程序，可以把字节代码编译成运行时的机器代码。

4．Java applet

Java applet 是以 Java 字节代码格式提供给用户的一种小应用程序。Java applets 可以用在使用 Java 虚拟机（JVM）的 Web 浏览器上；而在 Sun 的 Applet Viewer 中，它是一个独立的测试小应用程序的工具。

Java applets 的运行速度与其他编译语言，如 C++，有可比性，但速度比 C++等语言慢，但到 2011 年前后，它仍比 JavaScript 快很多倍。此外，Java applets 可以使用 Java 中的 3D 硬件加速功能。这使得 applets 很适合于最重要的、敏感的可视化计算任务。当浏览器获得了以 Canvas 和 Web GL，以及 Just in Time compiled JavaScript 格式表示的，本地硬件加速图形的支持时，其速度的差别便不明显了。

由于 Java 的字节代码是交叉平台或与平台无关的，因此 Java applets 可以用浏览器在很多平台上运行，包括微软 Windows、UNIX、Mac OS 和 Linux。通常 Java applet 可以与其他特别小的程序配合在一起应用。这种方式的优点是可以在离线方式下运行 Java applets，而无须因特网浏览器软件，并且可以直接使用集成开发环境（IDE）。

6.3 标记和脚本语言

有一些不是程序设计语言，而是与应用开发相结合的语言。这些语言大多数与万维网有关，如下面几部分所讨论的。

1. HTML 和其他标记语言

（1）HTML

现今大多数 Web 网页是用标记语言写的。采用标记语言，能用最少的线路容量实现在网上传输文件。标记语言通过使用各种标记定义了 Web 网页的结构和轮廓，而不必传送有关 Web 网页外形的精确描述。最通用的制作 Web 网页的标记语言是 HTML（超文本标记语言）。HTML 使用 HTML 标记。

在制作 Web 网页时——使用字处理软件、文本编辑器或专用的 Web 站点开发程序——HTML 标记是被插入到该 Web 网页文本的相应位置上的。某些标记单独使用，而其他一些则成对使用。例如，将标记后的文本转换为粗体，直到出现为止。因此下面的 HTML 语句：

`<b> This text is bolded </b>`

在用大多数 Web 浏览器观看时，可看到如下的粗体效果：

This text is bolded

一种 Web 网页及其相应的 HTML 代码如图 6-2 所示，其中带有一些通用的 HTML 标记。

HTML5 是用于构建和展示万维网内容的，因特网标记语言的一种核心技术。到 2014 年 10 月，公布了 HTML5 万维网联盟（W3C）的 HTML 标准的最终版和第五版。而前一版 HTML4 是 1997 年发布的。

这一核心的目标是改进该语言以支持最新多媒体，同时保持它的易读性，并能够使计算机和一些设备（万维网浏览器，语法分析程序，等等）更一致地理解。HTML5 力图不仅包括有 HTML4，也包括 XHTML1 和 DOM Level2 HTML。

具体而言，HTML5 增加很多新的语法功能，包括新 <video>、<audio>和<canvas>元件，和可以集成可缩放矢量图形的内容（取代一般的<object>标签），以及使用 MathML 数学公式。这些功能可以很容易地在万维网上处理多媒体和图形内容，而无须求助于相应的插件和应用程序接口。其他的构建页面的新元件，如<main>、<section>、<article>、<header>、<footer>、<aside>、<nav> 和<figure>是为了更丰富文档的语义内容的。为了同一目标引入了新功能，但同时有些元件和功能被去掉。另一些元件，如<a>、<cite>和 <menu>被变更，被重新定义或标准化。APIs 和 DOM 不再是追加部分，而是 HTML5 规范的基础部分。HTML5 也为无效文档的处理在某些细节上做了规定,因此，语法错误将由所有符合规范的浏览器和其他用户代理统一处理。

（2）XML

XML 是一种标记语言，它定义了一组对文档编码的规则，其格式既可供人读，也可供机器读。它是由 W3C 的 XML1.0 规范和其他几个相关规范定义的，全部为免费开放标准。

XML 的设计目标强调跨因特网时的简洁,通用和可用性。它具有文本数据格式,通过单一编码，对不同的语言有很强的支持能力。虽然 XML 的设计集中在文档上，但可广泛用于任意数据结构，如用于为万维网服务的那些数据结构中。

在定义基于 XML 的语言时，有几个图表系统很有用，同时有很多应用编程接口已被开发出来，用于辅助处理 XML 数据。

（3）.NET

与 XML 非常紧密相关的是微软公司的.NET 技术,是用来增加个人计算与 Web 的融合度的。简单地说，.NET 是微软公司实现基于 XML Web 业务的平台。这些业务允许应用程序通过因特网进行通信和共享数据，而与所用的操作系统或编程语言无关。

2. 脚本语言

（1）JavaScript

HTML 原则上是为具有活动元素的 Web 网页的页面布局设计的，很像桌面出版程序是为打印页面布局而设计的。因此 HTML 只有少量的制作 Web 网页的工具，仅当用户观看这些页面时可以做修改,或者允许用户在屏幕上与 Web 网页交互,而不具备 DHTML 和新的 HTML 增强版的某些功能。如果你希望开发的网页有大量的动态内容,采用像 JavaScript 这样的脚本语言比较合适。这些语言可使你将程序命令或脚本直接写入 Web 网页的代码中，以增加动态内容。例如，JavaScript 脚本语言，经常用于实现在一个菜单项被点击后，能够显示其子菜单或新的图形，如图 6-3 所示。

JavaScript 最初是由网景公司开发的，能使 Web 网页的作者建立交互式 Web 站点。虽然 JavaScript 具有很多丰富的 Java 语言的特性和结构，但它还是独立开发的。当使用 JavaScript 时，必须认识到并非所有脚本命令对所有浏览器都适合。因此，要确保你的网站使用的 JavaScript 的重要性能不是专门针对某个浏览器的。

（2）PHP

PHP 是用于万维网开发的服务器端的脚本语言，但也可作为通用编程语言使用。到 2013 年 1 月，PHP 已经安装在超过 2.4 亿个 Web 网站上（占取样数据的 39%），和 210

万台万维网服务器上了。

　　PHP 代码可以和 HTML 代码简单地混合使用，也可以与各种模板引擎和万维网框架组合使用。PHP 代码通常由 PHP 解释程序处理，它是作为 Web 服务器本地模块，或作为可执行的通用网关接口（CGI）去实现的。在 PHP 代码被解释和执行以后，Web 服务器就把结果用所产生的网页的部分形式送给它的客户端。例如，PHP 代码会产生一个 Web 网页的 HTML 代码，一幅图像或某些其他数据。PHP 也包含一个命令行接口（CLI），可用于独立的图形应用。

　　正规的 PHP 解释程序，由 Zend 引擎强化，该程序是在 PHP 授权之下的免费软件。PHP 已被广泛移植，并可在几乎所有 Web 服务器上、几乎每种操作系统和平台上免费使用。

　　尽管它很通用，但直到 2014 年，还没有编写出 PHP 语言的规范或标准，而将正规的 PHP 解释程序视为事实标准。2014 年以来，PHP 规范的制订工作一直在进行。

　　也是在 2014 年，新的 PHP，命名为 PHP7 也在开发之中。

第 7 章　数据库

7.1　数据库的概念

　　人们经常需要快速检索大量数据。一家航空代理在电话上要能快速为客户查到从亚特兰大到多伦多的最低价格机票。一个大学的注册主任可能要快速扫描学生记录，以查出将于 6 月份毕业的学绩点平均在 3.5 分或以上的那些学生。视频商店的店员可能需要决定某一种电影是否适合于零售。用于这些任务的软件就是数据库管理系统。计算机化的数据库管理系统正在迅速取代纸面上的文件系统，因为在过去，使用纸面上的文件系统，人们要费很大力气才能找到所需的信息。下面就讨论基于 PC 的关系型数据库的基本特性和概念，以微软的 Access 作为应用的例子。

1. 什么是数据库程序

　　数据库是按一定方式存储和组织数据的一种数据的集合，在需要时能够检索信息。数据库管理系统——有时也称为数据库软件——能够在计算机上创建数据库，并能很容易地访问存储在数据库中的数据。

　　虽然不是所有数据库的组织方式都一样，但大多数基于 PC 的数据库都是按字段、记录、文件去组织的。数据库中的字段是单一类型存储的数据，如一个人的姓名或电话号码。记录是相关字段的集合——例如，费利斯·哈弗曼的身份证号、姓名、地址和行业（见图 7-1）。文件，在 PC 数据库中常称为表——是相关记录的集合（如所有学生的地址数据、年级数据或课程表数据）。相关文件或表（如全体学生数据）的最终集合构成了数据库。

　　大多数 PC 的数据库软件是一种关系型数据库管理系统。

2. 建立一个数据库

数据库可以包含各种对象（见图 7-2）。在新数据库中最初建立的对象就是一张表，当需要时，可以建立其他对象，并与这个表连接。

在建立一个数据库时，应该确定要包含在数据库中表的数量。应该能识别存储在每张表中的数据项，以便使用相应字段的特性。对于每个字段，应该确定以下内容：

（1）字段名（表中的唯一识别名）

（2）要包含在字段中的数据类型（文本，数字，日期等）

（3）字段大小（存储数据需要多少个字符）

一旦这些技术要求确定下来，包含这些字段技术要求的每张表的结构就建立起来了（见图 7-3）。

表结构建立好以后，就可以往表里输入数据了。数据输入可以在常规表视图中完成，也可以建立和使用表单，常规表视图有时称为数据单视图，因为这种表看起来很像一张电子表格。表单可以使你以更正规的方法查看或编辑表中的内容——通常一次操作一个记录，而不是一页记录，如同在数据单视图中一样。图 7-3 说明，一旦表结构建立起来了，就可以使用这两种方法去输入数据了。

3. 修改数据库

一旦建立了数据库表，该表可能需要修改。当需要时可以对表的结构或放在表中的数据进行修改。

（1）修改表结构

只是当需要更改字段的性质时，才去修改表的结构。例如，可能要加宽一个字段，以容纳比以前设定的更长的名字，而错误的字段类型可能是当初选择的，也可能需要增加一个新字段。

（2）编辑、添加和删除记录

要更改表中的具体数据，应先把表打开（使用表的数据单视图或表单），然后对需要更改的数据做修改。为了移动到一个具体记录上，以便编辑它的内容，可以使用屏幕上的箭头和键盘上的其他定向键，也可以使用视窗底部的记录按钮（再请参见图 7-3）。因为通常是在表的末尾添加记录，在一组记录按钮中通常有一个新记录按钮（New Record），此按钮会自动移到表的末尾，成为一个空记录。

要删除一个记录，可以使用键盘上的删除（Delete）键或者使用菜单栏中的删除记录（Delete Record）选项。

4. 查询和报告

为了从数据库中检索信息，要使用查询和报告程序。查询是一种提问，在数据库术语中，查询是请求从数据库中找到指定的信息。查询程序用你希望查什么信息的指令，去查找并显示刚刚查到的信息。报告则是更正规地打印出一张表或查询结果。

7.2 万维网与数据库

在万维网上使用数据库是特别普通的。实际上，通过 Web 网站提供产品、合作信息、在线订货或类似活动的所有公司，都使用数据库。最常用的是客户-服务器数据库事务处理方式，在这种情况下，用户的浏览器为客户软件。然而，对等方式信息交换的使用也在不断增加。

1. Web 数据库使用举例

如何在 Web 上使用数据库有很多例子。数据库促进了信息的检索和处理，并允许更多的交互和动态内容的检索和处理。下面这几部分讨论的是 Web 数据库工作的例子，并扼要介绍一下其他与 Web 数据库相关的问题。

（1）信息检索

在 Web 上，用租用的数据库本身去进行数据检索是极其自然的，实质上，它是用作检索的特大数据仓库。数据存储在数据库中，Web 网站上的访问者可以请求和观看这些数据（见图 7-4）。

（2）电子商务和电子企业

另一种在 Web 上广泛使用的数据库应用，是支持并促进了电子商务的软件。目录信息、价目、客户信息、购物卡余额和其他信息都可存储在数据库中，当需要时可使用相应的与 Web 网站数据库连接的脚本文件或程序进行检索（仍见图 7-4）。

（3）动态 Web 网页

静态 Web 网页，每次在为个人显示时，所显示的是同一个信息，直到该网页文件被修改为止。相反，动态 Web 网页的外形和内容是按用户的输入变化的。这种输入可以由该网页上规定的表单选项来决定，也可以用其他方法控制，如 Java applet、ActiveX 控件，或用户已经完成的对该网站的一些动作，如单击一个显示的内容，特别是单击产品的超链接。

2. Web 数据库如何工作

（1）有关数据库和 Web 共同工作的例子

为进一步说明数据库和 Web 如何共同工作，让我们看一个例子。

向 Web 数据库请求检索信息或向该数据库存储数据通常由用户提出。实现数据库请求的通用方法是填写 Web 网页表单，从显示在 Web 网页上的菜单中选择一个选项，或单击屏幕上的项目。该请求由 Web 服务器接收，然后将此请求转换成数据库查询，并通过称为中间件的中间软件的帮助，将其传送到数据库服务器上。该数据库服务器检索出相应的信息并将其返回给 Web 服务器（要再次通过中间件），此 Web 服务器将所检索的信息，作为 Web 网页显示在用户的屏幕上，这些步骤表示在图 7-5 上。

（2）中间件

连接两个完全不同的应用程序——如图 7-5 所示的 Web 服务器和数据库管理系统——的软件称为中间件。用于数据库和 Web 网页之间接口的，最常用的中间件是 CGI 和 API 脚本程序。而更通用的、比较新的脚本语言是 PHP 和 ASP。

第四部分 应用软件

第 8 章 办公自动化软件

8.1 办公自动化软件基本知识

当你使用任何一种应用软件，如使用字处理软件打印一封信，或为纳税而使用税务处理程序时，一些基本概念和功能需要熟悉。其中包括通用文档处理任务、软件套件和软件所有权的概念以及当程序工作时如何得到帮助等。下面几部分就讨论这些问题。

1. 办公自动化软件基本概念

（1）文档处理操作

虽然某些文档处理操作是专为某一具体应用程序设计的，但像打开一个文档、存储和打印文档的概念是完全通用的。图 8-1 描述了少量最普通的文档处理操作，给出了在 Windows 应用程序中实现这些操作所使用的图标。

一般来讲，完成这些操作的命令在所有的 GUI 程序中是相同或非常相似的，因为绝大多数人需要快速使用文档处理命令，故这些操作通常放在容易单击的菜单中或工具栏上。

（2）软件套件

大多数办公用的程序，如字处理程序和电子表格程序，是与其他相关的应用软件以软件套件形式，捆绑在一起出售的。在办公应用套件的销售中，微软 Microsoft Office 稳坐第一把交椅。这一套件的高端版本捆绑了 Word（字处理）、Excel（电子表格）、PowerPoint（演示文稿）、Access（数据库管理），并且还与其他几个程序，如 FrontPage（Web 网站开发）一起出售。

使用软件套件的最大优点之一是能够把文档或部分文档从一个程序中传送到另一个程序中，或共享文档。例如，你正在用字处理程序写一封信，但是希望插入一张电子表格，你可以启动电子表格程序，在存储的表单中找到你需要的那张表格，把它复制并粘贴到你的信中就可以了——整个过程无须关闭字处理程序（见图 8-2）。

（3）在线帮助

大多数人在使用软件程序时会遇到问题，或在某一方面需要帮助。为了提供帮助而又不使你离开计算机屏幕，大多数应用程序有在线帮助功能。程序有各式各样的工具提供在线帮助。某些可能的配置如图 8-3 所示。

（4）所有权和分配权

涉及软件产品的所有权和用户权时是很敏感的。通常一个软件的制作者或发布者开发了一个程序，就在其上面制作一个版本权，并在以后保持这一程序的全部权限的所有权。以后由发布者决定谁可以使用、复制或分发该程序。下面讨论不同类型的所有权和允许使用权。

① 专有软件。

当今所用的很多系统软件和应用程序都是专有软件。这表明某人拥有该程序的权限，并且拥有者希望用户购买这些程序的复制品。

② 共享软件。

某些软件可作为共享软件使用。虽然你不必为安装和试验共享软件付费，但大多数共享软件规定在很短期限——通常为一个月——期限到了以后，你要继续使用该软件，则需要付费。

③ 免费软件。

免费软件，或称公用域软件，是指你可以使用并和其他人共享的免费程序。

2. 办公自动化软件的一般特性

用户接口是你使用的应用程序的一部分。大多数应用程序使用图形用户接口（GUI）去显示称为图标的图形元素，图标代表熟悉的对象和鼠标。鼠标控制屏幕上的指针，用于选择像图标那样的一些项。另一个特性是显示信息窗口的使用。窗口是可以包含文档、程序或消息的矩形区域（不要把窗口这一术语与微软公司的各种版本的 Windows 操作系统相混淆，那些是程序）。在计算机屏幕上一次可以打开和显示多个窗口。

包括微软 Office 2010 在内的大多数软件程序都有菜单、对话框、工具栏和按钮（见图 8-4）。菜单给出命令，通常显示在屏幕顶部的菜单栏内。当选中菜单中的一项时，会出现菜单选项列表或出现附加信息和请求用户输入的对话框。工具栏一般是在菜单栏下，它们包含称为按钮的小图形元素，提供快速访问常用命令的快捷方式。

微软最新的 Office 2010 版又重新设计了界面，旨在使用户更容易找到和使用应用软件的所有功能。这种新设计推出了条形框、上下文选项卡和图库等。

（1）条形框取代了菜单和工具栏，是把常用的命令组成一套选项卡。这些选项卡显示与用户正在处理的任务最相关的一些命令按钮。

（2）上下文选项卡是自动出现的一些选项卡。这些选项卡只是在需要时才出现，并指出用户要完成的下一次操作。

（3）图形库简化了从图形目录中选择的过程。这是由于某些结果采用图形显示而不是对话框之故。

8.2 微软 Office 2013

微软 Office 2013（以前是 Office 15）是微软办公软件的一个版本，是微软 Windows 高效率软件系列。它是微软 Office 2010 的后继者，包括扩展的文件格式，用户界面更新和这些新特性中的触摸功能。Office 2013 适合于 IA-32 和 x64 系统，要求配备 Windows 7、Windows Server 2008 R2 或以后的任何版本软件。Office 2013 版也可用在 Windows RT 设备上。

微软 Office 2013 有 12 种不同的版本，包括 3 个用于零售市场，两个用于批量授权，

5个通过微软 Office 365 程序使用的基于订阅的版本。还有称为 Office Web apps 的万维网应用版和为平板电脑与移动设备制作的 Office RT 版。

2014 年 2 月 25 日，微软 Office 2013 Service Pack1（SP1）面世。

1．新特性

比起以前的版本，Office 2013 更多地基于云；一个域登录，Office 365 账号或微软账号，现在可用于在设备之间同步 Office 应用软件的设定（包括最新的文档），用户也可以把文档直接存在他们的 SkyDrive 账号中。

新特性包括在微软 Word 中的新的读方式，在微软 PowerPoint 中的展示方式和所有 Office 程序中的触摸和喷墨方式。微软 Word 可以插入从在线源那里获得的视频和音频，也可以在万维网上广播文档。Word 和 PowerPoint 也有类似的书签功能，可在不同计算机之间同步文档的位置。

Office Web Apps 簇也对 Office 2013 做了更新，引入了附加的编辑功能和界面的更改。

Office 2013 的其他功能包括：

- 当敲键或进行选择时,会出现漂亮的带状界面和精巧的动画（Word 和 Excel 中）。
- 在 Outlook 中新的调度任务是可视化的。
- 改型的开始屏幕。
- Word 中的新图形选项。
- 诸如图像这样的对象，可以随意移动。这些图像可以很快地放到段落边缘、文档边缘或列的边缘。
- 从 Office.com、Bing.com 和 Flickr 网站来的内容，可支持在线图形（默认的只有公共域中的图像）。
- 在 Word 和 PowerPoint 中，可以返回到最后观看和编辑的位置。
- 在 PowerPoint 2013 中有新的幻灯设计、动画和变换。
- 在 Outlook 中支持 Outlook.com 和 Hotmail.com。
- 支持与 Skype、Yammer 和 SkyDrive 的整合。
- 支持 IMAP 专用文件夹。

图 8-5 为 Microsoft Office 2013 成员图标。

2．版本

与以前的各种版本一样，Office 2013 做成了几个不同的版本，以适应不同的市场需求。所有微软 Office 2013 的传统版本都包含 Word、Excel、PowerPoint 和 One Note，并都授权在一台计算机上使用。

公布的 Office 2013 的 5 个传统版本是：

- 家庭与学生版：这一零售系列包括 Word、Excel、PowerPoint 和 OneNote 的核心应用。
- 家庭与商业版：这一零售系列包括 Word、Excel、PowerPoint 和 OneNote+Outlook 的核心应用。
- 标准版：这一系列只能通过批量许可通道使用，包括 Word、Excel、PowerPoint 和

OneNote+Outlook 和 Publisher 的核心应用。
- 专业版：这一零售系列包括 Word、Excel、PowerPoint 和 OneNote+Outlook、Publisher 和 Access 的核心应用。
- 专业+版：这种系列只能通过批量许可通道使用，包括 Word、Excel、PowerPoint 和 OneNote+Outlook、Publisher、Access、InfoPath 和 Lync 的核心应用。

第 9 章 多媒体

9.1 多媒体及其主要特点

1. 定义

多媒体是指用各种形式的内容综合成的内容（它是相对于仅仅用基本的计算机文本显示、传统的打印形式或手写材料的媒体而言的）。多媒体是由文字、音频、静态图像、动画、视频或交互的内容形式的一种综合，如图 9-1 所示。

多媒体可以通过计算机和电子设备这样的信息处理设备进行记录、播放、显示、动态交互或进行访问，当然也可以是现场演出的一个部分。多媒体设备是用于存储和体验多媒体内容的电子媒体设备。多媒体在艺术方面区别于混合媒体就在于它包含了音频，因此，它具有更广泛的领域。"富媒体"一词与交互式多媒体是同义的。超媒体在多媒体应用中更加扩大了媒体内容的数量。

2. 多媒体分类

多媒体可以宽泛地分为线性媒体和非线性媒体。线性媒体，其活动过程通常对观众不进行导航控制，比如，演电影。非线性媒体则有交互控制过程，比如视频游戏或自定进程的基于计算机的培训。超媒体就是一个非线性媒体的例子。

3. 多媒体的主要特性

观众可以在舞台上观看多媒体演示，可以通过投影仪或传播后观看，多媒体也可以通过媒体播放设备在本地播放。多媒体播出可以是直播，也可以是录制的多媒体节目。广播和录制既可以采用模拟电子媒体技术，也可以采用数字电子媒体技术。数字在线多媒体可以下载，也可以按流式播出。流式多媒体既可以是直播，也可以是按需播出的。

通过多用户在线网络或本地的离线计算机、游戏系统或模拟器，多媒体游戏和模拟可用于有特殊效果的具体环境。

各种各样的技术手段或数字多媒体可以增强用户的体验，比如，更方便更快捷地传递信息。而在娱乐或艺术领域方面可以超越日常的体验。

将多种形式的媒体内容综合起来可以增强交互的水平。在线多媒体越来越多地变成面向对象和数据驱动的，这就能经过一段时间，使多种内容形式的终端用户的协作创新和个性化的应用成为可能。这方面的例子比如，从网站上多种形式的内容，如具有图像（相片）和用户更新标题（文字）的图片库到对其事件、插图、动画或视频的模拟有修改的可能，允许修改多媒体"体验"而无须再编程。除了视听之外，触觉技术还可以感知虚拟物体。新出现的包括味觉和嗅觉在内的技术，也可以增强多媒体体验。

9.2 多媒体应用

多媒体应用遍布各个领域,包括但不仅仅局限于:广告、艺术、教育、娱乐、工程、医疗、数学、商业、科研和时空应用领域。图 9-2 是使用 PowerPoint 做的演示画面。图 9-3 是使用多媒体内容的虚拟现实。图 9-4 是德累斯顿(德国)的一个多媒体终端。以下是一些应用例子。

1. 创意产业

创意产业从美艺术,到娱乐,到商业艺术,到新闻,到媒体各个领域都在使用多媒体,并且为下述任何行业提供软件服务。一个多媒体设计师在其职业生涯中可能涉及所有的相关领域,对他们的技能要求可以从技术,到分析,到创意。

2. 商业应用

商业艺术家和图形设计师使用的大量过往和新的电子媒体都是多媒体。动人心弦的展示用于在广告中抓住眼球。商业对商业之间,办公室之间的交流通常使用创新服务公司提供的先进的多媒体演示,而不是用简单的幻灯片去呈现新理念或吸引人的培训项目。商业多媒体开发者也被聘来设计政务服务和公益服务中的应用项目。

3. 娱乐和美艺术

此外,多媒体还大量地应用于娱乐业,特别是在电影和动画片中使用特效(VFX、3D 动画等)。多媒体游戏是流行的娱乐消遣方式,是以 CD-ROM 或在线方式发布的软件程序。一些视频游戏也使用多媒体功能。让用户积极地参与而非被动地只坐在那里接收信息的多媒体应用,被称为交互式多媒体。在艺术界,有一些多媒体艺术家,他们的想法是将不同媒体技巧进行混合来让观众以某种方式介入交互。

4. 教育

在教育方面,多媒体用于创作基于计算机的培训课程(常常称之为计算机辅助训练)以及制作像百科全书和年鉴这样的参考书。计算机辅助训练可以用各种信息格式让用户浏览一系列演示、某个主题的文本以及相关插图。寓教于乐就是教育和娱乐的结合,特别是在多媒体娱乐方面。

在过去的十年间,有关的学习理论由于引入多媒体而被梦幻般地扩展。研究的各个方向均被涉及(例如,认知负荷、多媒体学习及其他)。学与教的发展潜力几乎永无终点。

5. 新闻

所有的报业集团都试图在其工作中投向这一新领域。当一些报纸还缓慢前行时,其他那些像《纽约时报》《今日美国》和《华盛顿邮报》等主流报纸已尝试确立在报业全球化过程中的领先地位。

新闻报道已经不仅仅局限于传统媒体,自由新闻工作者可以通过不同的新媒体为他们的新闻故事制作多媒体报道。这也使得利用技术手段来实现听讲故事的全球化,也为媒体制作者和消费者提供新的交流技术。

6. 工程

软件工程可以使用多媒体对从娱乐到培训（如军事或工业培训）中的任何事件做计算机模拟。软件界面中的多媒体通常是创意人才和工程师们有机结合的产物。

7. 工业

在工业领域，多媒体是作为对投资者、上级和合伙人呈现信息的一种手段。在全球范围内，通过几乎无限的基于万维网的技术，多媒体也对员工培训、广告宣传和产品销售帮助良多。

8. 数学和科学研究

在数学和科学研究领域，多媒体主要应用于建模和模拟。比如，一个科学家可以观察一个特殊物质的分子模型并操纵它得到一个新的物质。

9. 医疗

在医疗领域，可以通过观看虚拟外科手术来训练医生，也可以模拟一个人体如何受病毒和细菌的传播而感染疾病的，从而开发新技术去防止这种感染。像虚拟手术这样的多媒体应用也可以帮助医生进行实操训练。

第 10 章　计算机图形图像

10.1　各种各样的计算机图形

1. 二维计算机图形

二维计算机图形是基于计算机产生的数字图像的——主要来自于模型，比如数字图像，以及通过特定的技术产生的。

最初，二维计算机图形主要用在从传统打印和绘图技术，例如排版，发展而来的应用领域中。在这些应用中，二维图像不仅是一个现实世界物体的表达，而且可以是添加了语义值的独立人造物体；二维模型因此更有优势，因为它们对图像的直接控制比三维计算机图形更多，其方式更像是摄影而不是排版。

（1）像素画

大量数字艺术品都是用光栅图像软件创建的像素图，其中的图像都是可以在像素级上进行编辑的。在多数老式（或相对有局限）计算机和视频游戏、图形计算器游戏和许多手机游戏中的图形基本上都是像素画。

（2）精灵图

精灵图是集成在一个大场景中的二维图像或动画。最初包含的只是一些图形对象，这些对象可单独在视频显示的内存位图中做处理，现在则包括各种各样的图形覆盖方法。

精灵最初是一个集成不关联位图，使之成为屏幕上普通位图的一部分的一个方法，例如创建一个动画角色，使其可以在屏幕上移动而不改变整个屏幕的数据。这种精灵既可以由电子电路也可以由软件来建立。用电路的话，硬件精灵是一个用客户 DMA 通道在主屏幕显示中集成的可视元素，从而超强加入了两个分离的视频源的硬件结构。用软件的话，则可以通过特殊的渲染方法来进行模拟。

（3）矢量图形

矢量图格式是对光栅图的补充。光栅图是以像素阵列来表示图像的，并且通常是用来表示图片的。矢量图是由构成图像的相关图形和色彩的编码信息组成的，它在渲染时有更多的灵活性。有很多例证可以说明使用矢量工具和格式更好，当然，也有很多例证可以说明使用光栅工具和格式更好。我们也常将这两种格式结合起来使用。对于每种技术的优点和局限性以及他们之间的关系就像是使用工具产生的效率和效果一样。

图 10-1 为矢量图与光栅图的对比效果。

2. 三维计算机图形

与二维图形相比，三维图形是用三维几何数据来表达图形的。它是为了展现而存于计算机中，它包含将来用于显示和实时观看的图像。

尽管有一些不同，3D 计算机图形依靠与 2D 计算机图形同样的算法在最终的渲染显示中做成帧和光栅图形（与在 2D 中一样）。在计算机图形软件中，2D 和 3D 的界限有时也模糊不清；2D 应用也使用 3D 技术来实现比如照明的效果，3D 也基本上使用 2D 渲染技术。

3D 计算机图形与 3D 模型一样。除了渲染之外，模型是包含在图形数据文件中。然而，不同之处在于包含的 3D 模型是任意 3D 对象的表达。直到可视化显示之前，一个模型并非一个图形。对于打印而言，3D 模型则不仅仅局限于虚拟的表面。3D 渲染就是如何对一个模型进行显示，当然也可用于非图形计算机模拟和计算。

3. 计算机动画

计算机动画是通过计算机创建移动图像的艺术。它是计算机图形和动画的分支领域。它越来越多地通过 3D 计算机图形进行创建，尽管 2D 计算机图形也依然广泛地满足低带宽的和快速的实时渲染需求。有时动画的目标就是计算机本身，但有时目标则是其他媒体，比如电影。尤其是用于电影时，它也被称为 CGI（计算机生成的图像或计算机成像）。

图 10-2 是使用 Motion 捕捉软件制作的计算机动画实例。

10.2 图形软件（1）

1. 桌面出版

虽然桌面出版（DTP）作为一项独立业务已有很长一段时间了，但它却是从文字处理发展起来的。近年来，这二者又显示出重新融为一体的趋势。基本的文字处理（WP）和 DTP 的不同之处可通过对传统印刷品作者所做的工作以及印刷工人所做的工作之间的比较来进行了解。在桌面计算机出现之前，作者必须产生一个打字稿——以正确顺序将文字排成文本。然后由印刷工人将这些文字（也许在设计人员或排字人员的辅助下）以特定的方式排好，其间也许有相应的插图，然后进行印刷。现在的作者与其前辈的做法一样，不同的是使用了文字处理软件，因而文字无须由印刷工人重排。DTP 所做的工作就是用桌面电脑来自动完成印刷工人所要完成的大部分工作。这一切来自于 4 项技术的发展：具有图形用户界面（GUI）的桌面电脑、DTP 软件、激光打印机和页面描述语言 PDL。

激光打印机的重要性在于可以获得高质量的最终文件，而无须传统的排字过程（特别是使用活字）。早期的 300~400dpi 的激光打印机还不能与传统印刷相匹敌，但已能以低廉的成本满足日常需要。尽管 DTP 在某些人的心目中仍然是低质量、业余水平的，但高分辨率激光打印机和数字排字（称为激光照排）的发展已可以用来制作出几乎任意质量的产品。

GUI 的重要性在于它能使 DTP 软件去展示预处理的文本和图形，并且 GUI 使用户能立刻看到准确的最终输出结果的样式。这个"所见即所得"（WYSIWYG）对 DTP 来说是极其重要的，尽管这句话似乎有些保留，因为所见的 72dpi 屏显不可能准确地反映你在 300~1200dpi（或更高）的打印机上所得到的效果。这句话也许该改成 WYSIANATTCMTWYG——"所见近乎技术上能做到之所得"。因此，DTP 软件包的一个必备特征就是有缩放工具，它可以用比正常情况更大的尺寸显示文档的某一部分，从而以更接近最终印刷结果的分辨率来显示这些内容。但遗憾的是在这种放大模式下，任一时刻只能查看文档的一小部分。

2. 电子出版（CD-ROMs 和因特网）

近年来，越来越多的资料是用电子的方式而不是在纸张上出版的。两个最重要的新媒体就是 CD-ROM 和 Internet。其结果就是许多 DTP 软件包和文字处理软件的新版本，都提供了专门为这两种媒体开发的格式进行文件输出的功能——特别是 HTML（超文本标记语言）和 PDF（可移植文档格式）。

超文本原理已被扩展至链接图片、图形、声音、视频、动画、图表、地图等。给用户带来的好处是：用户可得到大量信息，并且用户可以决定访问的量以及顺序。而印刷出版仍是一个线性过程，读者是以作者规定的顺序从头到尾进行阅读的，超链接信息的头和尾则都不是唯一的。

所有这一切表明，尽管还有许多与纸张出版物明显的相同之处，但电子出版是需要不同技术和不同软件的一个全然不同的媒体。

人们已设计出新型的不同种类的面向图形的软件，以生产各种电子出版物，如网上出版软件、网上图形设计软件、多媒体"制作"软件、PDF 出版软件。在研制此类新软件产品时所面临的一个潜在问题就是标准太多。已提到的两个标准之一就是 PDF 格式——是由 Adobe 公司通过其"Acrobat"软件套件建立的。另一个是 HTML 格式，目前正被扩展成 XML、DHTML 和 VRML。

10.3 图形软件（2）

1. 计算机辅助设计（CAD）

CAD 是图形应用的经典范例，它是在功能强大的大型计算机上发展起来的，并且人们曾经认为用桌面电脑完成此类重要任务是很不切合实际的。不错，现在许多重要的 CAD 工作仍然由图形工作站来完成，但越来越多的功能强大的应用系统也在普通桌面计算机上运行。最简单的计算机辅助设计之一就是二维平面制图——这种工作传统的方式是绘图员用丁字尺和圆规在绘图板上进行的。这类工作现在可以很快地由提供诸多工具的软件来完成。这些工具包括：电子丁字尺、橡皮擦、标准零部件和几何成形、示例、栅格、图层和标准图形库。

功能更强大的 CAD 软件包可用来设计和表现三维物体，这些软件包有时被称为"实体模型制造者"，因为它们在计算机上创建了三维物体完整清晰的模型描述，并可以用隐藏线消除和外形阴影来显示图像。由于三维建模涉及面很广，已远远超出了 CAD 的范畴。

2．绘图（画图）软件

虽然 CAD 软件与工程设计和其他形式的技术绘图相关，但这些软件包所提供的功能在许多其他应用中也被证明是非常有用的。可以认为绘图（或简称画图）软件是一个没有专门目的的 CAD 软件。像 CAD 软件一样，所绘制的图形可以以矢量方式制作和存储，因此，这类软件应和"绘图"软件区分开来，绘图软件通常用来创建和修改位图图像。CAD 和绘图软件的区别在于侧重点和方向的不同而非原理的不同。在这两种软件中，用户都可以对基于线条的几何图形进行组合和编辑；对象可以被单独选择、移动、变形及缩放；有边界的图形可用颜色和图案来进行填充；多个对象可以组合并看作是单个物体，都提供像栅格和快照之类的多种定位辅助手段。典型的用户可以用这种软件来产生较专业的绘图，同时也可以用它制作非专业的广告设计、宣传册和杂志，因此，其区别在于需要很多强大的功能，而不需要太多专业性。通常这些应用需要很多字体、颜色和彩色效果、现成的、全然不同的图形库和更复杂的曲线。这种图形软件包生成的图像会少一些实用性而多一些美感——更像是"艺术品"。

3．商业演示软件

这是另一种具有明确应用范围的矢量软件——快速生成用于尽可能紧凑地总结一些概念或数据的"演示"（报告或短文）。主要的特点是：简单的绘图功能（像图例说明软件那样）、剪辑功能、具有吸引力的演示、少量文字和图表。与绘图软件或 CAD 的用户不同，此处的用户对象并非设计人员，因此能自动生成具有美感效果的作品是十分重要的。这一点是通过各式各样的"模板"来实现的。通常，模板是用来帮助人们建立某种对象的模式或导引——就像用刻有字母的塑料模板快速地在纸上描字一样。在图形软件中，模板是一种设计好的东西，用户可以用它快速地完成自己的任务。比如，它可以是一屏"虚化"的文本——用户只须用自己的文本进行替换即可。关键在于专业的图形设计师已设计好模板——选择好了字体和颜色，或许还有图案背景——用户无需任何设计技巧就可生成专业水平的效果，其速度也比从头做起快得多。

第 11 章 现代工业自动化

11.1 CAD、CAM、CAE 的应用

上面已经描述了一个典型的产品周期，下面来看一下在此周期内如何使用计算机或

CAD、CAM 和 CAE 技术。正如前面所指出的，在设计过程的综合阶段，计算机用得并不广泛。这是因为计算机不能很好地处理定性信息。但是，在综合子过程中，比如，设计人员可以通过商业数据库很好地为可行性研究搜集相关的设计信息，并以同样的方法搜集目录信息。

同样很难想象计算机如何应用于设计概念化阶段，因为目前计算机还谈不上是智能创造过程的强大工具。在这一阶段，计算机有助于高效率地完成各式各样的概念设计。参数建模、计算机辅助绘图的宏编程能力或几何建模，在完成此类任务时会很有用。这些软件包是 CAD 软件的典型例子。可以想象出来，一个几何模型系统相当于一个三维绘图系统，也就是说，通过这种软件包就可以用三维形体来取代二维图片。

设计过程中的分析子过程最能体现计算机的价值。实际上，对于应力分析、干扰检验和运动分析有许多现成的软件包。这些软件包归入 CAE 一类。使用这些软件包的一个问题就在于如何提供分析模型。如果能从概念设计中自动导出分析模型的话，那就根本不是问题。但是，正如前面说过的，分析模型与概念设计不同，它是通过从设计中忽略了不必要的细节或缩小尺寸而得来的。根据分析模型的类型和所期望求解精度的不同，就有不同的抽象水平。因此，很难对抽象过程进行自动处理；而分析模型通常是分别建立的。常用的做法是，用计算机辅助绘图系统或几何建模系统或有时使用分析软件包的内置功能，为设计过程多建几个抽象模型。

分析子过程可以嵌入到优化循环中以获取最优设计。获取最优解决方案的各种算法都已开发出来，并且许多优化程序亦可在市场上购买到。优化过程可以认为是 CAD 软件的一个部分，但最好还是单独对待。

设计评估阶段使用计算机也会受益匪浅。如果为了设计评价而需要一个设计原型的话，则可以用自动生成驱动原型速成机的软件包对一个给定的设计构建一个原型。这些软件包都归类为 CAM 软件。当然，要做的原型形状应事先以某种数据类型存在。对应于该形状的数据应该由几何建模系统来创建。尽管原型可以方便地用原型速成方法来构建，但如果使用虚拟原型就会更加理想。虚拟原型通常称为数字模型，它提供同样有价值的信息。

随着用于评价数字模型的分析工具的功能变得足够强大，其给出的分析结果与在真实原型上相同实验所获得的结果同样精确时，数字模型就有了取代真实原型的趋势。随着虚拟现实技术能够使数字模型获得与真实原型相同的效果，这种趋势将不断增强。建立数字模型的工作称为虚拟原型设计。虚拟原型也可由专为此设计的某类几何建模系统来产生。

设计过程的最后阶段是设计文档。在这一阶段，计算机辅助绘图是一个强大的工具。计算机绘图系统的文件处理能力也允许对文档进行系统地存储和检索。

计算机技术也同样可用于制造过程。制造过程包括：生产计划、新工具的设计和采购、材料订购、数控编程、质量控制和包装等活动，因此应用于这些活动的计算机技术都属 CAM。例如，用于辅助工艺计划的计算机辅助工艺计划（CAPP）软件就是 CAM 软件的一种。前面曾提到过，工艺计划很难自动完成，因此 100%自动化的 CAPP 软件目前还拿不到。但是，有很多优秀的软件包能够产生驱动数控设备的数控程序。当某种形状以数据形式存在于计算机中时，此类设备就可以加工出给定的形状。这类似于驱动原型速成机。此外，同属 CAM 的还有：对机器人运动进行编程，以便按零部件进行组装或将其送至各个制造活动中的软件包，或对坐标测量机（CMM）编程，以便对产品进行检查的软件包。

11.2 3D 打印

3D 打印（或称增材制造）是用来制作三维物体的各种加工技术。在 3D 打印中，使用增材加工技术，在这种技术中，在计算机控制下，对材料的连续层进行铺设。这些物体可以是各种形状或几何结构的，并能从 3D 模型中产生，或从其他电子数据源产生。3D 打印机是一种工业机器人。

图 11-1 为一种 3D 打印机。

按原始术语的意思，3D 打印是指用喷墨打印头，连续往粉末床上沉积材料。该术语最新的意思已扩展到包含更为广泛的技术，诸如基于工艺的挤压和黏结等。从更广的意义上讲，技术标准应采用增材制造术语。

1. 基本原理

（1）建模

可以用 CAD 软件包，或 3D 扫描仪，或普通的数码相机和摄影测量软件去创建 3D 打印模型。

为 3D 计算机图形准备几何数据的人工建模过程，类似于造型美术，像雕刻一样。3D 扫描是对一个实体在形状和外观上采集和分析数字数据的过程。在这种数据的基础上，被扫描物体的三维模型就产生出来了。

与所使用的 3D 建模软件无关，3D 模型（通常用.skp、.3ds 或其他格式）然后要转换成 .STL 或 .OBJ 格式，以便打印（又称"CAM"）软件去读它。

图 11-2 为三维模型切片。

（2）打印

在从 STL 文件打印 3D 模型前，首先要检查"复印误差"，这一步可以称为"修复"。特别是通过 3D 扫描获取的模型所产生的 STL，通常有很多复印误差需要修正。例如，复印误差是表面的，则不需要在模型中去连接或补缺。可用于修正这些误差的软件有 netfabb 和 Meshmixer，甚至可使用 Cura 或 Slic3r。

一旦做完上述工作，.STL 文件需要一个称为"切片机"的软件去处理，该软件会把上述模型转换成一系列薄层，并产生一个 G-code 文件，此文件含有专为某种 3D 打印机（FDM 打印机）定制的一些指令。这种 G-code 文件可用 3D printing 客户软件打印（该软件装入 G-code，并在 3D 打印过程中指导 3D 打印机）。应该注意的是，实际上这种客户软件和切片机通常组成一个软件程序。现在有几种开源切片机程序，包括 Skeinforge、Slic3r 和 Cura-engine 以及闭源程序，包括 Simplify3D 和 KISSlicer。3D 打印机客户软件包括 Repetier-Host、ReplicatorG、Printrun 和 Cura。

（3）结束

虽然打印机打出的结果，分辨率能满足很多应用的要求，但在标准分辨率内打印时，其尺寸应稍微超过所需物体尺寸，这样，在其后采用更高分辨率的削减材料过程中，会获得更高的精度。

某些可以打印的聚合物，允许对表面做平滑处理，用化学蒸发工艺做进一步改进。

某些增材制造技术能够在构造零件过程中使用多种材料。这些技术能同时打印多种颜色和组合颜色，且不要求涂色。

所有商业化的金属 3D 打印机都包含，在沉积之后，把金属部件从金属衬底上切掉。一种新的用于 GMAW 3D 打印的工艺，可以对衬底表面做修正，用人工小锤去掉铝制部件。

2．应用

20 世纪 80 年代，增材制造在产品开发、数据可视化、快速成型和专用制造中开始应用。此后几十年，在开发中扩展到了生产（如制品生产、批量生产和分布式制造）。从 2010 年开始的几年间，工业生产首先在金属加工行业获得了长足进步。从 21 世纪开始，增材制造机器销售量大幅增长，而价格大体上是下降的。按照某个咨询公司 Wohlers 助理的说法，2012 年世界范围内，3D 打印机及服务的市场价值为 22 亿美元，比 2011 年增长 29%。增材制造技术有很多应用，包括建筑、施工、工业设计、自动化、航天、军事、工程、牙科和医药行业、生物技术（人组织置换）、服装、鞋业、珠宝饰品、眼睛饰品、教育、地理信息系统、食品和其他很多领域。

图 11-3 为表明工业中 3D 打印技术优越性的涡轮模型。

2005 年，随着开源 RepRap 课题的启动，3D 技术很快就有了广泛的嗜好者和家庭市场。实际上，迄今为止，所公布的所有家用 3D 打印机都在 RepRap 课题和相应开源软件项目中有他们的技术根基。在分布式制造中，一项研究表明，3D 打印会变成一个海量市场产品，能让客户在购买一般家用物品时，节省大量钱财。例如，可以不去商店购买工厂注入成型的产品（如量杯或漏斗），人们可以在家中，从下载的 3D 模型中将其打印出来。

BIBLIOGRAPHY

[1] Timothy J O'Leary. Linda I O'Leary. Computing Essentials 2014. 北京：机械工业出版社，2015.

[2] 刘艺，王春生. 计算机英语[M]. 4 版. 北京：机械工业出版社，2013.

[3] 司爱侠，张强华. 计算机英语教程[M]. 6 版. 北京：电子工业出版社，2014.

[4] https://en.wikipedia.org/wiki/Windows_10.

[5] https://en.wikipedia.org/wiki/Microsoft_Office_Mobile.

[6] https://en.wikipedia.org/wiki/MySQL.

[7] https://en.wikipedia.org/wiki/Wi-Fi.

[8] https://en.wikipedia.org/wiki/Bluetooth.

[9] https://en.wikipedia.org/wiki/Massive_open_online_course.

[10] https://en.wikipedia.org/wiki/Flipped_classroom.

[11] https://en.wikipedia.org/wiki/Big_data.

[12] https://en.wikipedia.org/wiki/Industry_4.0.

[13] https://en.wikipedia.org/wiki/Internet_of_Things.

[14] https://en.wikipedia.org/wiki/WeChat.

[15] https://en.wikipedia.org/wiki/Home_network.

[16] https://en.wikipedia.org/wiki/Storage_area_network.

[17] https://en.wikipedia.org/wiki/Internet_area_network.

[18] https://en.wikipedia.org/wiki/Twitter.

[19] https://en.wikipedia.org/wiki/Cloud_computing.

[20] https://en.wikipedia.org/wiki/Computer_graphics.

[21] https://en.wikipedia.org/wiki/Product_lifecycle.

[22] https://en.wikipedia.org/wiki/3D_printing.

[23] https://en.wikipedia.org/wiki/Multimedia.